Functional analysis in historical perspective

Functional analysis
in historical perspective

DR. A. F. MONNA

Professor of Mathematics

Rijksuniversiteit Utrecht

1973

A HALSTED PRESS BOOK

JOHN WILEY & SONS

New York-Toronto

Published in the USA and Canada by Halsted Press, a Division of John Wiley & Sons, Inc., New York.

Printed in the Netherlands

Library of Congress Cataloging in Publication Data
Monna, A. F.
Functional analysis in historical perspective.
'A Halsted Press book.'
Bibliography: p.
1. Functional analysis—History. I. Title.
QA320.M66 515'.7'09 73-14414

ISBN 0-470-61327-0

Contents

CONTENTS

Foreword

There are indications that there is a growing interest among mathematicians in the history of the development of mathematics. I believe that there is reason to be glad with such a tendency. After many years of explosive development it is worthwhile reflecting on the way mathematics is going, and therefore the study of the course of the development of the subjects with which mathematicians occupy themselves in our days is important. Many of the problems of contemporary mathematics find their roots in classical problems and many old problems are still alive, be it sometimes in a form in which it is difficult to recognize these old problems. It is worthful that any mathematician be aware of this fact.

The present book is written from this point of view. In contemporary mathematics functional analysis plays an important role. Concepts such as Hilbert space, normed space, Banach space are nowadays common tools. How did they enter mathematics? Origins lie in the 19th century in mechanical and physical problems. In the hands of famous mathematicians like Fredholm, Hilbert, Poincaré, Riesz, Schmidt and Banach the theory gradually developed into what we nowadays call functional analysis. Subjects connected with this development are the theory of infinite systems of linear equations with infinitely many unknowns and the theory of integral equations. There is a close connection with the introduction into mathematics of the concept of a linear space; the first steps in this direction were taken by Grassmann, Peano and Pincherle. This led to the penetration of the algebraic methods into analysis, which is an important feature not only of modern analysis, but of all presentday mathematics.

The present book provides a readable survey of this development. It is not a textbook on functional analysis; it contains a description of its gradual development. Profound knowledge of the subject is not required; any student with a certain amount of mathematical maturity will be able to read the book and it will give him a background for modern mathematics. It is believed that any mathematician can profit from it in view of his mathematical erudition.

I do not claim completeness and I have only slightly touched on the latest developments. In a full discussion one should be forced to explain many details. On the one hand this was impossible without considerably expanding the text and on the other hand it is perhaps not yet the right moment for a detailed discussion because functional analysis is still in a stage of development. So I propose to give a survey of the great lines. In advance I apologize for shortcomings from the point of view of the historian.

In order to simulate the readers to study themselves in the sources there are many quotations from the older literature in the text. An extensive bibliography will help those who want to penetrate into the subject in finding their way.

I thank my colleagues D. van Dalen, H. Freudenthal, T. A. Springer and J. van Tiel for their most valuable critical remarks and their assistance. My thanks go also to Miss W. Jenner for the excellent way in which she typed the manuscript.

SUMMER 1973

A. F. MONNA

Introduction

At the second international conference of mathematicians, held in Paris in 1900, the Italian mathematician Volterra characterized the 19th century as the century of the theory of functions. This conference is especially known by the fact that at this conference Hilbert gave his famous lecture in which he put his problems which should be solved in the 20th century. Hilbert was not the only mathematician who spoke at this conference on the trend of the development of mathematics: Maurice Cantor – not to be confused with Georg Cantor – gave a lecture on 'L'Historiographie des Mathématiques' and Vito Volterra's lecture was entitled: 'Betti, Brioschi, Casorati, trois analystes italiens et trois manières d'envisager les questions d'analyse'. It is in this lecture that Volterra spoke the words with which I began. Volterra's lecture seems not very well known, being in the shadow of Hilbert's lecture. Nevertheless it is worthwhile reading.

Volterra discussed the development of analysis in the 19th century by analysing the scientific work of the Italian mathematicians, named in the title of the lecture who had a great influence but were quite different in their attitudes towards mathematics. He gave the following characterization of analysis in this period ([125], p. 47, 48):

La théorie qui a eu le plus grand développement dans les derniers temps est sans doute la théorie des fonctions. On pourrait même appeler notre siècle, au point de vue des Mathématiques, le *siècle de la théorie des fonctions*, comme le XVIIe siècle pourrait être désigné par le nom de *siècle du calcul infinitésimal*.

Volterra observed that three periods can be distinguished in the development of the theory of functions. I quote from his lecture the following passage ([125], p. 48, 49):

Il existe bien des travaux historiques et critiques sur la théorie des fonctions. Les plus savants géomètres ont donné sur elle des essais précieux, riches des plus intéressantes notices.

I

Mais jetons un coup d'œil d'en haut sur le chemin parcouru, envisageant dans son ensemble le développement de la théorie.

Nous distinguons tout de suite trois phases différentes qui marquent presque trois périodes distinctes.

D'abord s'élaborent des théories particulières. C'est leur développement qui montre la nécessité de créer une théorie générale des fonctions transcendantes et des fonctions algébriques, qui embrasse tous les cas connus et en prévoit de nouveaux. Dans cette phase on ne connaît pas encore de méthodes uniformes. Chaque question qui se présente, on doit tâcher de la résoudre : voilà ce qui s'impose. Les méthodes, il faut les créer chaque fois et à chaque pas. De longs calculs sont nécessaires et les pensées qui sont renfermées dans les formules ne se dégagent que peu à peu.

Les grands noms d'Euler, de Jacobi, d'Abel peuvent être pris pour personnifier cette période héroïque où la théorie des fonctions elliptiques a été créée dans ses parties essentielles, et où ont été marquées les lignes principales où devaient se développer un jour les fonctions abéliennes.

Mais à cette période de découvertes merveilleuses, où ce qui domine est la curiosité d'arriver en possession de vérités inattendues qui se dévoilent soudainement à travers de longs calculs et des inductions audacieuses, succède bientôt une phase où l'esprit philosophique a le dessus et où s'impose la nécessité de la recherche d'une méthode générale et puissante qui embrasse et renferme tout dans un cadre unique en constituant un corps de doctrine.

Cette phase est marquée par les œuvres immortelles de Cauchy, de Weierstrass et de Riemann qui sont remontés aux sources mêmes des conceptions fondamentales pour accomplir leur tâche. C'est dans cette période grandiose que les idées remplacent peu à peu les calculs.

Il y a enfin une dernière phase où les théories trouvent leurs plus importantes et leurs plus fécondes applications, les formes les plus appropriées à leur diffusion, et restent fixées dans un cadre didactique, après avoir été passées en revue et discutées par le plus fin esprit critique qui ait jamais dominé la Science.

Ces trois phases, dont nous avons tâché de donner les principaux caractères, correspondent à peu près à trois périodes successives dans l'histoire de la théorie des fonctions, mais elles correspondent aussi à trois manières d'envisager les questions d'Analyse; et certains géomètres restent attachés à l'une ou à l'autre en vertu même des qualités les plus intimes de leurs esprits.

Volterra showed that Brioschi is a representative of the first period, Betti of the second and Casorati of the third period by analysing their work.

What about *analysis* in the 20th century and the last years of the 19th century? It is a fascinating activity to look in the older literature for characteristic tendencies with regard to the development of analysis in this period. One of my aims is to show the continuity in the development of mathematics. A study in this domain should not be a chronology because a strict chronology is not suitable for describing the characteristics of a development. Furthermore the period which I have in view–I begin in the middle of the 19th century–is rather short for a chronological description.

In the present book I will try to describe some features in the development of analysis in such a way that the facts are ordered with respect to certain points of view. For example a comparison between the way of treating certain problems by our modern methods and by those which were used half a century ago; there are several examples to illustrate the difference.

Two aspects should be distinguished: on the one hand I mention the problems themselves and on the other hand there are the methods by which these problems are attacked and proved.

As regards the kind of problems: I have the intention to treat mainly the development of the *functional-analytic features* of modern analysis. Nowadays one sometimes discerns between *hard analysis* and *soft analysis*. Hard analysis more or less corresponds to what is called *classical analysis*; it is not the point of view which I take in this essay. I am going to treat soft analysis; the meaning of the term 'soft analysis' will become clear in the course of this book.

However, hard and soft analysis are not strictly separated. The tools used in the proofs of properties in hard analysis are nowadays different from those that were used half a century ago. Functional-analytic methods in hard analysis are common in modern analysis.

3

I mention for instance the theory of differential equations and integral equations where methods are used belonging to the theory of Hilbert space and, more generally to the theory of normed spaces or even locally convex spaces. In his book *Théorie des opérations linéaires* (1932) Banach used, for instance, functional-analytic methods to prove the existence of continuous nowhere differentiable functions.

I propose to sketch some features of the development of functional analysis by treating some characteristic examples which show that this part of mathematics gradually developed from classical mathematics and physics. I don't aim at completeness; rather I think that the examples clearly indicate the course of the development.

It seems to me that functional-analytic methods are an outstanding feature – I don't want to say the only characteristic feature – of modern analysis. Functional analysis exhibits connections between parts of mathematics which at first sight seem to be alien. This will be explained in the course of the book.

Seen against the background of analysis as it was half a century ago, I believe three aspects are dominant. They can shortly be described as follows: *1* a tendency towards algebraization; *2* a stream towards results of a structural character; *3* the strong influence of topology.

I will give a short explanation at this point although the meaning of these three points will become clearer in the course of the essay.

As to the points *1* and *3* an explanation seems scarcely necessary. In our century there is a strong penetration of algebra into analysis. Results from classical analysis are generalized in such a way that they have acquired a mixed character of algebra and analysis. I think, for example, of the modern version of the classical theorem of Weierstrass concerning the approximation of continuous functions by polynomials. I mention the theory of function algebras and of ideals in these algebras. This is what I call the algebraisation.

The strong influence of topology needs no comment: topology is basic for fundamental notions such as limit and continuity. I mention for instance the different types of convergence: strong and weak convergence.

What I mean by the second point – structural proporties – will become clear in the following pages. But I can give a rough sketch. In modern

4

analysis–at least in those parts of analysis that I have in mind, although the same tendency can be found in hard analysis–results on families of functions (spaces) considered as a whole are important, in some parts even dominant, whereas the properties of the individual functions of the family are not of primary interest. I mention the theory of function spaces which has turned out to be important for classical theories such as the theory of partial differential equations, treated nowadays with the theory of distributions. We observe here the transition from a function to a *class* (*a space*) of functions where the individual properties of the elements of the class are not a subject of study and where, on the other hand, one or more structures on the class are defined, for example a group structure, a field structure or a topological structure, connected with each other in some adequate way. Ultimately this development has led to an abstract analysis in the form of an axiomatic theory in which more 'concrete' theories are brought under the same point of view.

I am not going to treat these aspects strictly separated; this would be impossible–at least undesirable–because of the connections between them.

One of the things that will strike the reader of modern books and articles on analysis is the use of the words *space, function space, linear function space* or, in the axiomatic setting, *normed space, Banach space*. This is a geometric terminology, but there is no need at all to have a geometric picture of these concepts; the spaces are mostly infinite dimensional. On the other hand a geometry in these spaces was developed, sometimes with properties analogous to properties in euclidean geometry.

How did spaces and especially linear space get introduced in analysis? There are various ways which led to spaces in analysis, all starting in the end of the 19th century. I mention the Italian school with Pincherle, Volterra ('fonctions de ligne'), Peano. Then Hilbert, Schmidt, F. Riesz, Helly, Hahn, Banach. Then we are at ± 1920. In the beginning the word 'space' was not yet used, this notion developed gradually.

Families of sequences of real numbers, satisfying some conditions, were studied; the 'totalité des fonctions continues' and more 'concrete' spaces. Abstract spaces were introduced in a later stage, where one should have to mention *Fréchet*–with his 'analyse générale'–who based his work on the Italian school. My way of dealing with these aspects will not be chronological; this does not seem very relevant.

It is important to distinguish between the introduction of the algebraic notion of space (linear algebra), connected with the names of *Grassmann* and *Peano*, and the introduction of topologies on these spaces, leading to metric spaces, normed spaces.

I will treat three sources: *1* infinite systems of linear equations with an infinite number of unknowns; *2* integral equations; *3* the problem of moments.

The first two originate from physics and mechanics, the third from the theory of probability. There are close connections between these sources, especially between the first two (by means of the theory of orthogonal systems of functions). Yet I will try to treat these sources separately as far as possible; the connection will appear in the course of the essay.

CHAPTER I

The development of functional analysis

It seems that infinite systems of linear equations appear for the first time with Fourier. In his famous *Théorie analytique de la chaleur* (1822) he treats a problem in the theory of partial differential equations which brings him to consider a system of countably many linear equations. Fourier finds a solution of his system, be it with methods that cannot stand the test of criticism. In his book from 1913 *Les systèmes d'équations à une infinité d'inconnus* [109] F. Riesz states that this example of Fourier had been virtually forgotten for more than half a century and that the mathematicians who came to study these infinite systems were unaware of this first result of Fourier. I quote the example here because I suppose that nowadays it is forgotten again. In the presentation I follow Riesz l.c.

Problem. One asks for a solution of the partial differential equation

$$\Delta v = v''_{xx} + v''_{yy} = 0$$

in the domain $x > 0$, $-\frac{1}{2}\pi < y < \frac{1}{2}\pi$, which equals 1 for $x = 0$ and vanishing for $y = \frac{1}{2}\pi$ and $y = -\frac{1}{2}\pi$, and for $x = \infty$.[1]

Fourier puts

$$v(x, y) = F(x)f(y)$$

and he determines F and f without regarding the boundary values. This leads to a solution

$$v(x, y) = e^{-mx} \cos my$$

of the equation for any real value of m. For $m = 1, 3, 5, ..., 2m-1, ...$

---

[1] Fourier's formulation of the last condition is ([27], p. 163): 'Il faut ajouter que cette fonction $\varphi(x, y)$ doit devenir extrêmement petite lorsqu'on donne à x une valeur très grande, puisque toute la chaleur sort du seul foyer A.'

the boundary conditions except the first one are satisfied. In order to satisfy also this condition Fourier considers the series[1]

$$v(x, y) = \sum_{m=1}^{\infty} a_m e^{-(2m-1)x} \cos(2m-1)y,$$

in which the coefficients a_m are to be determined in such a way that

$$1 = \sum_{m=1}^{\infty} a_m \cos(2m-1)y$$

for all $-\frac{1}{2}\pi < y < \frac{1}{2}\pi$. In order to eliminate y Fourier differentiates the series an infinite number of times term by term and puts $y = 0$. This leads to the equations

$$1 = \sum_{1}^{\infty} a_m,$$

$$0 = \sum_{1}^{\infty} (2m-1)^2 a_m,$$

$$0 = \sum_{1}^{\infty} (2m-1)^4 a_m,$$

$$\cdot \cdot \cdot \cdot \cdot \cdot \cdot \cdot \cdot \cdot \cdot$$

This is a system of infinitely many linear equations for the a_m.

To find a solution of this system Fourier only considered $a_1, ..., a_k$ and the first k equations, neglecting the terms with $a_{k+1}, a_{k+2}, ...$. An elementary calculation provides the corresponding values of $a_1^{(k)}, a_2^{(k)}, ..., a_k^{(k)}$. He then calculated

$$\lim_{k \to \infty} a_m^{(k)}.$$

Taking for a_m the value of this limit, he finds

$$v(x, y) = \frac{1}{4}\pi \sum_{m=1}^{\infty} (-1)^{m-1} \frac{e^{-(2m-1)x} \cos(2m-1)y}{2m-1},$$

and

---

[1] Fourier did not use the notation with indices; he wrote

$$v = ae^{-x} \cos y + be^{-3x} \cos 3y + ce^{-5x} \cos 5y + de^{-7x} \cos 7y + \&,$$

as was customary in those years (l.c. p. 164).

$$\tfrac{1}{4}\pi = \cos y - \tfrac{1}{3}\cos 3y + \tfrac{1}{5}\cos 5y - \dots,$$

for $-\tfrac{1}{2}\pi < y < \tfrac{1}{2}\pi$.

Now, as Riesz remarks, there are serious objections to the way in which Fourier obtains this result. He differentiates the series I mentioned above term by term without any scruples whether this is allowed. And indeed, when substituting the values calculated for a_m in the linear system, it appears that these series from the second one on are divergent. I let aside the question whether Fourier was aware of the objections to his heuristic method. I only indicate that, having found the series for $\tfrac{1}{4}\pi$, he proves directly that the sum of this series is constant for $-\tfrac{1}{2}\pi < y < \tfrac{1}{2}\pi$ and equals $\tfrac{1}{4}\pi$. Moreover he states that the series found for v, satisfies $\Delta v = 0$ and has the desired boundary values. One may guess that Fourier found it necessary to verify his results directly. For some more information on the results of Fourier I refer the reader to Riesz l.c.

But apart from these objections, the method of Fourier was very important with regard to the general principles for solving infinite systems of the form

$$\sum_{k=1}^{\infty} a_{ik}x_k = c_i, \quad i = 1, 2, \dots.$$

Riesz l.c. p. 7 writes as follows:

Voici ce principe, qui bien entendu devra encore être beaucoup précisé: Pour résoudre un système infini d'équations à une infinité d'inconnues, on limite le système aux k premières équations et l'on y néglige toutes les inconnues, sauf les k premières. Les solutions de ces systèmes tendent, pour k infini, vers la solution du système proposé.

Ce principe, dont la légitimité peut être justifiée sous des hypothèses très large, se montrerait très fécond pour la théorie en vue. Nous l'appellerons *principe des réduites*.

This principle, which one can regard as a passage 'from finite to infinite', played an important role in the development of the theory of infinite linear systems. Somewhat analogous is a principle in the theory of integral equations, where it can be regarded as 'from discrete to continuous'; I will return to this matter. This method preceded the way by which in a later stage these problems were attacked directly, that is to say without limiting processes; these later theories led to the theory of linear operators.

The following example, due to Riesz l.c., p. 8, is a nice illustration of the danger hidden in the application of the 'principe des réduites'.

Example. We consider here the theory of analytic functions of a complex variable. Let $(a_i)_{i \in N}$ be a sequence of complex numbers such that $|a_i| \to \infty$. It is well known that there exists an entire function f vanishing for $z = a_i$ and only for these values of z:

$$f(z) = c_0 + c_1 z + c_2 z^2 + \dots.$$

The coefficients c_i are the solutions of the infinite system

$$c_0 + c_1 a_i + c_2 a_i^2 + \dots = 0, \quad i = 1, 2, \dots.$$

Suppose that $a_i \neq 0$ $(i \in N)$ and $a_i \neq a_j$ $(i \neq j)$. Now, consider the product

$$\prod_{i=1}^{n} \left(1 - \frac{z}{a_i}\right) = c_0^{(n)} + c_1^{(n)} z + \dots + c_n^{(n)} z^n, \quad (c_0^{(n)} = 1).$$

For $n = 1, 2, \dots$ these products correspond evidently to the equations that one obtains by the 'principe des réduites'. The solutions are therefore $c_0^{(n)}, \dots, c_n^{(n)}$. According to this principle one must consider the limits $\lim_{n \to \infty} c_i^{(n)}$ and ask whether these limits are the solutions of the infinite system. This is in general not true. Indeed, the product converges for $n \to \infty$ only under certain restrictions on the a_i. One concludes that the infinite system of linear equations always has a solution but only under certain restrictions this solution can be obtained as the limit for $n \to \infty$ of the solutions of the reduced equations. As is well known from the theory of functions of a complex variable the product must be multiplied by some factors in order to be convergent.

Another curious example is due to *Helly* [65]. He considered the infinite system

$$x_1 + x_2 + x_3 + \dots = 1$$
$$x_2 + x_3 + \dots = 1$$
$$x_3 + \dots = 1$$
$$\dots \dots$$

Successively subtracting the equations one sees that $x_i = 0$ ($i = 1, 2, \ldots$) but this is not a solution of the system. This system has no solution at all. However, any of the reduced finite systems has a solution. The 'infinite determinant' of the system equals 1, if it is defined as the limit of the finite determinants. The solvability of the system depends on the members of the right side: the system

$$x_1 + x_2 + x_3 + \ldots = 1$$
$$x_2 + x_3 + \ldots = 0$$
$$x_3 + \ldots = 0$$
$$\cdots\cdots$$

has the solution $x_1 = 1$, $x_i = 0$ ($i > 1$) and now it is the limit of the solutions of the reduced systems.

These simple examples show the complexity of the theory of infinite systems of linear equations.

After Fourier, infinite systems of linear equations were not studied for about half a century. About 1870 some mathematicians again considered such systems, without reference to Fourier. Fürstenau (1860) was led to these systems in studying some algebraic problems concerning the roots of an algebraic equation. A result in this domain, due to Von Koch, related to the problem of Fürstenau, is the following:

Let f be an entire function of the complex variable:

$$f(z) = 1 + c_1 z + c_2 z^2 + \ldots.$$

Let α be a root of the equation $f(z) = 0$ with least absolute value. Let $(x_i)_{i \in N}$ be a solution of the system

$$0 = x_0 + c_1 x_1 + c_2 x_2 + c_3 x_3 + \ldots$$
$$0 = \qquad x_1 + c_1 x_2 + c_2 x_3 + c_3 x_4 + \ldots$$
$$\cdots\cdots\cdots$$

Then

$$\limsup_{n \to \infty} |x_n|^{1/n} \geqq |\alpha|.$$

Kötteritzsch (1870) studied such systems in a more general case. As is quite natural by the analogy of the theory of finite systems there are some hints to the theory of 'infinite determinants'; but it is evident that for infinite systems conditions on convergence will be indispensable. For details on these beginnings of the general theory I refer the reader to Riesz [109] from which I have taken these indications and where one finds a bibliography.

The decisive steps were set by Hill, Poincaré, Von Koch and Hilbert in the years round 1900. The origin was a problem in celestial mechanics. In studying the motion of the moon Hill [67] came to consider the differential equation

$$\frac{\mathrm{d}^2 u}{\mathrm{d}t^2} + \theta u = 0,$$

where

$$\theta = \theta_0 + 2\theta_1 \cos t + 2\theta_2 \cos 2t + \dots,$$

that is, if $\theta_{-n} = \theta_n$,

$$\theta = \sum_{-\infty}^{+\infty} \theta_n e^{int}.$$

Hill tries to put u in the form

$$u = \sum_{-\infty}^{+\infty} \lambda_n e^{i(n+c)t},$$

where the constants λ_n and c must be determined in such a way that u is a solution. This leads to the doubly infinite system of linear equations

$$\sum_{k=-\infty}^{+\infty} \theta_{n-k}\lambda_k - (n+c)^2 \lambda_n = 0, \quad -\infty < n < \infty.$$

This system is more complicated than the former ones because the corresponding matrix is infinite in four directions:

$$\begin{bmatrix}
\vdots & \vdots & \vdots & \vdots & \vdots & \vdots \\
\dots \theta_1 & \theta_0 - c^2 & \theta_{-1} & \theta_{-2} & \theta_{-3} & \theta_{-4} \dots \\
\dots \theta_2 & \theta_1 & \theta_0 - (1+c)^2 & \theta_{-1} & \theta_{-2} & \theta_{-3} \dots \\
\dots \theta_3 & \theta_2 & \theta_1 & \theta_0 - (2+c)^2 & \theta_{-1} & \theta_{-2} \dots \\
\vdots & \vdots & \vdots & \vdots & \vdots & \vdots
\end{bmatrix}$$

Hill gives a solution of the system applying classical methods (quotient of determinants) where he introduces infinite determinants and operates with them as if they were finite, however without verifying the legitimacy of his procedures. Hill's results were in agreement with the observations.

Poincaré [102] completes Hill's work. Then, under the influence of Mittag-Leffler, Von Koch made his contributions to the theory of infinite determinants. He defines the determinant of a doubly infinite matrix as follows ([74], [75]):

Soit A_{ik} $(i, k = -\infty \dots + \infty)$ une suite doublement infinie de nombres donnés et désignons par

$$D_m = [A_{ik}]_{i, k = -m \dots + m}$$

le déterminant des quantités A_{ik} $(i, k = -m \dots + m)$; si, pour des valeurs indéfiniment croissantes de m, la quantité D_m a une limite déterminée D, on dit que le déterminant infini

$$[A_{ik}]_{i, k = -\infty \dots + \infty}$$

est convergent et a D pour valeur.

The theory of the infinite systems

$$\sum_{k=1}^{\infty} a_{ik} x_k = c_i, \quad i = 1, 2, \dots,$$

was thus developed in close connection with the theory of infinite determinants.[1] It is evident that conditions on the a_{ik} and c_i will be necessary in order to obtain a solution. Riesz ([109] p. 42) states this as follows:

Pour appliquer la méthode classique des déterminants aux systèmes

---

1 For recent literature on infinite determinants see a paper of Sikorski [115].

infinis, il fallait imposer aux données des conditions plus ou moins restrictives, et il faut avouer que c'est la méthode et non le problème qui exigeait ces restrictions. Dans ce qui suit, nous nous placerons à un point de vue beaucoup plus général, dû en principe à M. E. Schmidt. Nous commencerons par préciser la nature des solutions admises (x_k) et, quant aux données, nous ferons la seule hypothèse que pour tout système admis (x_k), les séries figurant au premier membre des équations doivent être convergentes. La question principale sera celle de l'*existence* des solutions; il s'agira seulement en second lieu de l'appareil qui les fournit.

It is clear that, in order to show the existence of a solution, it will be necessary to agree on what is to be understood by a solution of the system. It is not a pure algebraic problem; statements on convergence are indispensable. Nowadays we would say: we must know in which linear space, i.e. a linear space of sequences, we want to have a solution. The conditions on the coefficients of the equations then express conditions on continuity or some other restrictions of the corresponding transformations in this space. But in the beginning of the development of the theory the problem was not formulated in terms of spaces, although the notion of a linear space was known, especially in the Italian school (Volterra, Peano, Pincherle); but it was not commonly used. I will treat the introduction of vector spaces in chapter II.

Riesz states in his book from 1913 [109] the following problem (but in his earlier publications one already finds similar problems).

Consider the system

$$\sum_{k=1}^{\infty} a_{ik}x_k = c_i, \quad i = 1, 2, \ldots.$$

Let be given a number $p > 1$. One wants a solution (x_k) such that the series

$$\sum_{k=1}^{\infty} |x_k|^p$$

is convergent.

For the left hand side of the equations to be convergent for any system (x_k) which satisfies this condition it is necessary and sufficient (Hellinger, Toeplitz, Landau) that

$$\sum_{k=1}^{\infty} |a_{ik}|^{1/(p-1)}$$

is convergent for any i. Using this, Riesz then obtained the following result:

There is a solution such that

$$\sum |x_k|^p \leqq M^p$$

if and only if

$$|\sum_{i=1}^{n} \mu_i c_i| \leqq M (\sum_{k=1}^{\infty} |\sum_{i=1}^{n} \mu_i a_{ik}|^{p/(p-1)})^{(p-1)/p}$$

for all values of n and for all real numbers μ_i.

This is proved directly, that is to say by a method which avoids the detour of considering the system as the limit for $n \to \infty$ of a finite algebraic system and without the use of infinite determinants. Riesz used a certain property, which he called 'principe de choix'; the meaning of this principle is a kind of compactness (I will return to this matter in § 3). Hölder's inequality [68] and an inequality which is a consequence of it, now called the triangular inequality, are fundamental for the proof:

$$|\sum_{k=1}^{\infty} a_k b_k| \leqq (\sum_{k=1}^{\infty} |a_k|^{p/(p-1)})^{(p-1)/p} (\sum_{k=1}^{\infty} |b_k|^p)^{1/p}, \quad (p > 1)$$

$$(\sum_{k=1}^{\infty} |a_k + b_k|^p)^{1/p} \leqq (\sum_{k=1}^{\infty} |a_k|^p)^{1/p} + (\sum_{k=1}^{\infty} |b_k|^p)^{1/p}.$$

For $p = 2$ the first inequality goes back to Cauchy.

Nowadays we say that we consider the infinite system in the normed space l^p and the meaning of the exponents $p/(p-1)$ and $(p-1)/p$ is quite clear since, when $q = p/(p-1)$, one has $1/p + 1/q = 1$ and l^q is the dual space of l^p (for the definition of l^p see p. 36). The theorem is now a simple consequence of the theorem of Hahn-Banach, but in those years this theorem was not yet known. Conversely the theorem of Hahn-Banach gradually developed from these special cases.

For $p = 2$ one gets a theorem in *Hilbert space*; but, as I said before, in the beginning one did not use this terminology, not even the word space. In the years from 1904 till 1910 Hilbert published very fundamental work on integral equations leading to Hilbert space, which he connected with infinite systems of linear systems; see § 2. Hilbert himself did not use the term 'space' in this context at that time. In 1913 Riesz used the name 'espace de Hilbert' in his book [89]; on page 78 he wrote:

> Considérons l'*espace Hilbertien*; nous y entendons l'ensemble des systèmes (x_k) tels que $\sum |x_k|^2$ converge. Nous étudierons les substitutions linéaires à une infinité de variables, portant sur l'espace Hilbertien.

And on page 73 of this book, in a passage where he refers to the work of E. Schmidt, one finds:

> Interprétons les systèmes (a_k) ou (x_k) lorsque $\sum a_k^2$ ou $\sum x_k^2$ converge, comme des vecteurs dans l'espace à une infinité de dimensions. Etant donné un nombre fini ou une infinité de vecteurs, les vecteurs orthogonaux à chacun d'eux constituent une certaine variété linéaire, et l'ensemble des vecteurs orthogonaux à cette dernière variété sera la plus petite variété linéaire contenant les vecteurs donnés. Par suite, quand il s'agit de résoudre un système homogène, on n'aura qu'à envisager les coefficients comme des coordonnées de vecteurs, la variété orthogonale à ces vecteurs représentera la totalité des solutions. Les systèmes non homogènes peuvent être rendus homogènes en introduisant une inconnue auxiliaire.

Infinite dimensional vector spaces and in particular Hilbert space had by then become an object of study in analysis and one realised that a geometry was possible in these spaces with some analogy to the geometry in n-dimensional euclidean spaces. However, we will see later on that infinite dimensional spaces appeared earlier than in the year in which Riesz wrote this book. I shall have occasion to return to it more than once.

What is the geometrical interpretation of the infinite system of linear equations in these sequence spaces? It is evident that they define linear transformations in the underlying spaces. From 1904 on the theory of orthogonal transformations in an infinite number of variables, in connection with integral equations, was a fundamental research subject for Hilbert. I will treat this theory in § 2. In Riesz's book [109] (1913) there is a chapter entitled 'Théorie des substitutions linéaires à une infinité de variables'. In modern mathematics we use the denomination 'linear operators'. Thus the theory of the infinite systems of linear equations became one of the starting-points of the theory of these operators. This was done in the frame of the theory of linear spaces but, as I said before, the theory of these spaces is much older; I will treat this matter in chapter II. In the theory of these spaces one is concerned with the properties of sets of sequences or functions, considered in their totality, not with properties of the individuals.

I have to mention here some important contributions to the theory of infinite linear systems. First I mention E. Schmidt; I mentioned him before (c.f. the quotation of Riesz p. 14). He wrote the important paper 'Über die Auflösung linearer Gleichungen mit unendlich vielen Unbekannten' (1908) [113].

Schmidt built on the work of Von Koch and Poincaré but especially he was inspired by the very important work of Hilbert on integral equations. The way in which Schmidt treats the problem differs essentially from the methods of his predecessors. He proceeds *directly*, that is he avoids the way of considering the infinite system as the limiting case of a finite system. He does so by using geometric methods which he develops in order to apply them to infinite systems. He considers for instance a homogeneous system (in the notation of Schmidt):

$$a_{n1}Z_1 + a_{n2}Z_2 + \ldots + a_{nm}Z_m + \ldots \text{ ad inf} = 0, \quad n = 1, 2, \ldots \text{ ad inf}$$

and he supposes that $\sum_{m=1}^{\infty} a_{nm}^2 < \infty$ for $n = 1, 2, \ldots$. He wants a solution such that

$$\sum_{m=1}^{\infty} |Z_m|^2 < \infty.$$

This means that one has to determine all vectors $(Z_1, Z_2, \ldots)$ which are orthogonal to the linear space, generated by the vectors $(a_{n1}, a_{n2}, \ldots)$ for $n = 1, 2, \ldots$; I remind the reader of the well known inner product of two vectors.

Inhomogeneous equations are treated in an analogous way. Explicit formulae for the solutions are given. I won't give the details of Schmidt's methods; for my purpose it is important that Schmidt in

1908 developed a geometry in some kinds of function spaces. Some examples follow below. The first chapter is entitled: 'Geometrie in einem Funktionenraum' and the first paragraph in this chapter 'Der pythagoräische Lehrsatz und die Besselsche Ungleichung'. The function space which Schmidt considers is of a very special type: it is in fact a space of sequences, this space is useful for treating infinite systems of linear equations. He introduces these function spaces in the following way:

Mit grossen Buchstaben, die vor das eingeklammerte x gesetzt werden, wie z.B. mit $C(x)$ sollen in dieser Untersuchung durchweg Funktionen von folgenden Eigenschaften verstanden werden.

I. *Die Funktion ist nur für $x = 1, 2, 3, \ldots$ ad inf definiert.*

II. *Die Quadratsumme der absoluten Beträge der von ihr durchlaufenen Werte convergiert.*[1]

He states that, if A and B satisfy these conditions, this is also the case for $\alpha A + \beta B$. He defines

$$(A; B) = (B; A) = \sum_{x=1}^{n=\infty} A(x)B(x),$$

and

$$\|A\|^2 = \sum_{x=1}^{\infty} |A(x)|^2.$$

He does not use the word 'norm' but he calls $A(x)$ 'normiert' if $\|A\| = 1$. Compare the later work of Helly (1921) (see [65]). Then A and B are orthogonal if

$$(A; \overline{B}) = (B; \overline{A}) = 0.$$

In a footnote Schmidt refers to some other kinds of function spaces, introduced by Fréchet (1906) and Fischer (1907). Some remarks in the Encyklopädie-paper of Hellinger and Toeplitz on integral equations and systems of linear equations with an infinite number of unknowns are noteworthy [63]. On page 1434 one finds the following reference to Schmidt:

Die Grundlage seiner (i.e. Schmidt's) Untersuchungen bilden gewisse allgemeine Begriffe der Geometrie des unendlich dimensiona-

---

1 Note that these functions are just the 'sequences' in modern notation.

len Raumes, die auch sonst von Bedeutung sind und hier zunächst in Zusammenhang dargestellt werden sollen.

The following footnote is added:

Diese sind zuerst von D. Hilbert (4. Mitteilung, Gött. Nachr. 1906 = Grundzüge, Kap. XI) aufgestellt worden, soweit er sie für seine Untersuchungen brauchte. E. Schmidt hat sie für seine Zwecke weiter ausgebildet, in der Darstellung aber die geometrische Form vermieden; er spricht von 'Funktionen eines ganzzahligen Index' statt von Punkten und Vektoren des unendlichdimensionalen Raumes. In geometrischer Sprache hat P. Nabholz [Dissert. Zürich 1910, 118 S. und Vierteljahrsschrift Zür. Naturf. Ges. 56 (1911), p. 149–155] die Schmidtschen Untersuchungen dargestellt.

It seems to me that the remark concerning Hilbert should not be understood literally. In the first place, in Hilbert's work that is quoted here he does not introduce any geometrical notions in infinite dimensional spaces. He does not even use the word space nor does he mention properties of linearity of the sets he considers. Hilbert treats the analysis in an infinite number of variables $x_1, x_2, \ldots$, restricted by an inequality like $\sum x_i^2 \leqq 1$, especially he treats orthogonal transformations. There are no indications in this book concerning an infinite-dimensional geometry.

From the available literature it appears that Schmidt was much closer to the idea of geometrical analogies than Hilbert. It is true indeed that Schmidt avoids the use of the words 'vector' and 'space' in the context, but a footnote shows that he was very well acquainted with these notions; there he writes ([113], p. 56):

Die geometrische Deutung der in diesem Kapitel entwickelten Begriffe und Theoreme verdanke ich Kowalewski. Sie tritt noch klarer hervor, wenn $A(x)$ statt als Funktion als Vector in einem Raum von unendlich vielen Dimensionen definiert wird. Die zugrunde gelegte Definition der Länge $\|A\|$ und der Orthogonalität sind die von Study [Math. Analen, Bd. LX, p. 372] in die Geometrie eingeführt.

Schmidt was, as he remarks, certainly inspired by Hilbert. I mention for instance his definition of strong convergence in the class of functions which he considers:

$$\lim_{n=\infty} D_n(x) = D(x)$$

stands for

$$\lim_{n=\infty} \|D - D_n\| = 0.$$

and he adds in a footnote:

> Die Unterscheidung zwischen gewöhnlicher Convergenz und starker Convergenz entspricht der Hilbert'schen Unterscheidung zwischen stetiger und vollstetiger Funktionen, l.c., S. 200.

Schmidt considers notions like 'Funktionenmenge' and 'Häufungsfunktion'. A function $A(x)$ is called 'Häufungsfunktion'[1] of a set $\mathcal{M}$ wenn es zu jeder positiven Grösse ε eine von $A(x)$ verschiedene Funktion $M(x)$ in $\mathcal{M}$ giebt sodass $\|A - \mathcal{U}\| < \varepsilon$ ist. Wenn $\mathcal{M}$ alle seine Häufungsfunktionen enthält, heiszt $\mathcal{M}$ abgeschlossen.

Linear spaces are introduced under the name 'lineare Funktionengebilde':

> Eine abgeschlossene Funktionenmenge soll dann ein *lineares Funktionengebilde* heissen, wenn aus der Zugehörigkeit von $A(x)$ und $B(x)$ zur Menge auch die Zugehörigkeit von $\alpha A(x) + \beta B(x)$ für alle Werte von α und β folgt.

These notions are not in Hilbert's *Grundzüge*, at least not explicitly.

I have to mention some other mathematicians. First E. Helly and his paper 'Über Systeme linearer Gleichungen mit unendlich vielen unbekannten' (1921) [65]. Helly was inspired by the so called 'Problem of moments' and by the theory of linear functionals with which I will deal in § 5. The theory of functionals originated from the Italian school at the end of the nineteenth century (Pincherle, Volterra) and was developed gradually by many mathematicians (Hadamard, Fréchet, F. Riesz). Therefore I postpone the discussion of Helly in the frame of the development of the theory of functionals. Helly's work must be considered an anticipation on the important theorem of Hahn-Banach.

Then there is H. Hahn and the important paper 'Über lineare Gleichungssysteme in linearen Räumen' (1927) [54]. In this paper the theorem of Hahn-Banach, one of the fundamental theorems of

---

[1] Häufungsfunktion = cluster function = accumulation function.

functional analysis, is anticipated; the proper place to discuss this theorem is in § 5.

Of course I have not mentioned all papers on the domain of infinite systems of linear equations, of infinite matrices and determinants; I refer the reader to the paper of Hellinger and Toeplitz [63] and also their joint article on the same subject 'Grundlagen für eine Theorie der unendlichen Matrizen' [62]. Furthermore there is A. Wintner's *Spektraltheorie der unendlichen Matrizen* [137]; in this book the theory of Hilbert space is already one of the main tools.

Up to now I gave only some hints on the fundamental work of Hilbert on our domain. The reason is that his work can best be treated in connection with the theory of integral equations, which has many relations with the theory of infinite systems of linear equations and was another source for modern functional analysis.

§ 2 *Integral equations*

For the introduction of integral equations we have to go back in history again. They find their origin in the beginning of the 19th century and, just as in the case of Fourier's mathematical innovations, the integral equations owe their birth to physical and mechanical problems.

The Norwegian mathematician Niels Abel was the first one who studied an integral equation and succeeded in solving it. His results were published in the *Magazin for Naturvidenskaberne*, Aargang 1, Bind 2, Christiania, in an article entitled 'Solution de quelques problèmes à l'aide d'intégrales définies' [1]. The equation resulted from a problem in mechanics. It is the following problem:

C'est bien connu qu'on résout à l'aide d'intégrales définies, beaucoup de problèmes qui autrement ne peuvent point se résoudre, ou du moins sont très difficiles à traiter. Elles ont surtout été appliquées avec avantage à la solution de plusieurs problèmes difficiles de la mécanique, par exemple, à celui du mouvement d'une surface élastique, des problèmes de la théorie des ondes etc. Je vais en montrer une nouvelle application en résolvant le problème suivant.

Soit *CB* une ligne horizontale, *A* un point donné, *AB* perpendicu-

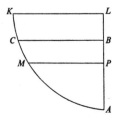

laire à BC, AM une courbe dont les coordonnées rectangulaires sont $AP = x$, $PM = y$. Soit de plus $AB = a$, $AM = s$. Si l'on conçoit maintenant qu'un corps se meut sur l'arc CA, la vitesse initiale étant nulle, le temps T qu'il emploie pour le parcourir dépendra de la forme de la courbe, et de a. Il s'agit de déterminer la courbe KCA pour que le temps T soit égal à une fonction donnée de a, p. ex. ψa.

Si l'on désigne par h la vitessse du corps au point M, et par t le temps qu'il emploie pour parcourir l'arc CM, on a comme on sait

$$h = \sqrt{BP} = \sqrt{a-x}, \; dt = -\frac{ds}{h},$$

donc

$$dt = -\frac{ds}{\sqrt{a-x}},$$

et en intégrant

$$t = -\int \frac{ds}{\sqrt{a-x}}.$$

Pour avoir T on doit prendre l'intégrale depuis $x = a$ jusqu'à $x = 0$, on a donc

$$T = \int_{x=0}^{x=a} \frac{ds}{\sqrt{a-x}}.$$

Or comme T est égal à ψa, l'équation devient

$$\psi a = \int_{x=0}^{x=a} \frac{ds}{\sqrt{a-x}}.$$

In modern language the problem can be stated as follows:

A function φ be given. If $y = y(x)$ is the equation of the desired curve, the arc length s of the curve is determined by

$$s'(x) = \left\{ 1 + \left(\frac{dy}{dx} \right)^2 \right\}^{\frac{1}{2}}.$$

One gets the equation

$$\sqrt{2g}\,\varphi(a) = \int\limits_0^a \frac{s(x)dx}{\sqrt{h-x}}.$$

This is an *integral equation* for the arc length s.

More generally Abel determines s from the equation (written down here in Abel's notation)

$$\psi a = \int\limits_{x=0}^{x=a} \frac{ds}{(a-x)^n}, \quad 0 < n < 1,$$

in which $ds^2 = dx^2 + dy^2$. He finds the solution by means of classical differential and integral calculus putting

$$s = \alpha^{(m')} x^{m'} + \alpha^{(m'')} x^{m''} + \alpha^{(m''')} n^{m'''} + \ldots.$$

Using the theory of Γ-functions[1] he finds – without scruples about the exactness of the proof – the solution

$$s = \frac{\sin n\pi}{\pi}\, x^n \int\limits_0^1 \frac{\psi(xt)dt}{(1-t)^{1-n}}.$$

As an application of this method Abel determines the equation of the isochronous curve by putting $\psi a = c$, in which c is a constant. This leads him to the cycloidal curve.

Abel's work had only little influence. There is a paper of Sonine in 1884 [116] which contains a generalization of the equation that was

---

[1] With respect to the Γ-function Abel l.c. remarks that 'Γm est une fonction déterminée par les équations $\Gamma(m+1) = m\Gamma m$ et $\Gamma(1) = 1$' and that 'Les propriétés de cette fonction ont été largement développées par M. Legendre.'

studied by Abel. There is a modern application of Abel's equation in Fourier analysis; see R. Godement, *Introduction aux travaux de A. Selberg*, Sém. Bourbaki no. 144 (1957), pp. 1–16. However, in the meantime there had been other progress.

In the thirties of the 19th century Liouville was led to study the same type of integral equations as a consequence of his investigations concerning certain series connected with the theory of differential equations; see [84].[1] From Volterra I take the following problem which was studied by Liouville (see [111]).

> Une droite indéfinie *y*, sur laquelle il y a une distribution de masses uniforme et symétrique par rapport à l'axe des *x*, est attirée par un point *A* situé sur cet axe à la distance *x*. L'attraction sur chaque point de *y* dépend de la distance au point *A*, mais la loi en est inconnue. L'attraction totale $\psi(x)$ étant donnée, déterminer l'action élémentaire $F(r)$ du point *A* sur un point *M* de la droite, situé à une distance *r* de *A*.

After some substitutions on ψ and F and putting $x^2 = z$, this problem leads to the integral equation

$$\varphi(x) = \int\limits_{z}^{\infty} \frac{u(\xi)\mathrm{d}\xi}{\sqrt{\xi - z}},$$

from which *u* has to be solved when the function φ is given. This equation was also solved by classical methods.

For more details on the integral equations in this preliminary period I refer the reader to the book of Volterra on integral equations [127].

There is another approach of integral equations that I want to mention. It is found in *potential theory*. Again the source is in physics. Potential theory is concerned with the solutions of the equation (which I consider in threedimensional space $\mathbb{R}^3$, but there are no essential differences in $\mathbb{R}^n$, $n > 3$; there is a slight difference in $\mathbb{R}^2$)

$$\Delta u = \frac{\partial^2 u}{\partial x^2} + \frac{\partial^2 u}{\partial y^2} + \frac{\partial^2 u}{\partial z^2} = 0.$$

[1] Volterra [127] observed that Liouville obtained his results independently of the work of Abel.

The solutions of this equation are called *harmonic functions*. There is a connection with potentials in physics: if μ is a measure (in the older books on potential theory this is called a distribution of mass) consider the potential F:

$$F(x) = \int \frac{d\mu_y}{\|x-y\|},$$

where $\|x-y\|$ is the euclidean distance of the points x and y. Then F is a harmonic function outside the support of the measure (outside the mass). The theory of these functions has a long history: it goes back to Gauss and earlier. Several famous mathematicians studied potential theory: Dirichlet, Riemann, C. Neumann.

An important problem in potential theory–perhaps the most important–is the following so called *Dirichlet's problem*:

Let Ω be an open set in $\mathbb{R}^3$ and let be given a continuous real function f on the foundary $\partial\Omega$ of Ω. The problem is whether there exists a function F, defined and continuous on the closure $\bar{\Omega}$ of Ω such that $f = F$ on $\partial\Omega$.

I will not treat the history of this famous problem which is, in a generalized form, today still a subject of research; I content myself by pointing out that there are sets Ω for which the problem has no solution for any f in this classical sense. I remind the reader of the discussions round the so called *principle of Dirichlet*, a method which was used in demonstrating the existence of a solution in some cases and whose exactness was only proved later by Hilbert. For all this see the classical book of Kellogg [71].

With regard to my purpose of describing the sources of functional analysis I have to mention here the reduction of Dirichlet's problem to an integral equation. Therefore I must require the boundary $\partial\Omega$ of Ω to be sufficiently smooth. Let us take $\partial\Omega$ to be a smooth surface; there is a normal in every point. Let λ be a positive measurable function on $\partial\Omega$. Consider the function $U: \Omega \to \mathbb{R}$ defined by

$$U(x) = \int_{\partial\Omega} \lambda(\xi) \frac{d}{dn} \frac{1}{\|x-\xi\|} \, d\omega,$$

where d/dn designs the normal derivative in the point $\xi \in \partial\Omega$, chosen in a suitable direction of the normal n. In classical potential theory U

25

is called the potential of the double layer on $\partial\Omega$ with momentum λ. One easily sees that U is harmonic in Ω. Now one tries to represent the desired harmonic function on Ω as the potential of a double layer for a suitably chosen function λ. Using well known relations between the values of U on $\partial\Omega$ and the limits of U when x tends to the boundary from the inside respectively from the outside of Ω, one finds an integral equation for λ:

$$f(\xi_0) = -\lambda(\xi_0) + \int_{\partial\Omega} \lambda(\xi) \frac{\mathrm{d}}{\mathrm{d}n} \frac{1}{\|\xi_0 - \xi\|} \,\mathrm{d}\omega_\xi.$$

The idea of this reduction is due to Beer (1865). He gave a solution, however not in an exact way. Under certain restrictions on Ω Carl Neumann succeeded in proving the convergence of the series by which he tried to represent the solution ('Neumann'sche Reihe') [93].

The equations which were considered are special cases of the following *general* integral equation:

$$\varphi(x) + \int_a^b K(x, y)\varphi(y)\mathrm{d}y = \psi(x), \quad (a \leqq x \leqq b)$$

in which ψ and K are given functions and φ is to be determined; in the equation for potential theory the *kernel* K is symmetric, i.e. $K(x, y) = K(y, x)$. It is called an equation of the *second kind*.

The equations of Abel and Liouville are of the type

$$\int_a^b K(x, y)\varphi(y)\mathrm{d}y = \psi(x).$$

They are called equations of the *first kind*.

Two methods of solution

By way of example I mention two procedures for solving an integral equation of the second kind.

A. *Successive approximation.* The following sequence of functions is formed:

$$\varphi_0 = \psi,$$

$$\varphi_1(x) = \psi(x) - \int_a^b K(x, y)\varphi_0(y)\mathrm{d}y,$$

$$\cdots\cdots\cdots\cdots\cdots\cdots$$

$$\varphi_n(x) = \psi(x) - \int_a^b K(x, y)\varphi_{n-1}(y)\mathrm{d}y, \quad n = 1, 2, \ldots.$$

If the sequence (φ_n) is uniformly convergent, the limit is shown to be a solution of the equation.

B. *Iteration.* The following sequence is considered:

$$\psi_1(x) = -\int_a^b K(x, y)\psi(t)\mathrm{d}t,$$

$$\psi_2(x) = -\int_a^b K(x, y)\psi_1(t)\mathrm{d}t,$$

$$\cdots\cdots\cdots\cdots\cdots\cdots$$

$$\psi_n(x) = -\int_a^b K(x, y)\psi_{n-1}(t)\mathrm{d}t, \quad n = 1, 2, \ldots.$$

Under some conditions $\psi + \sum_{i=1}^{\infty} \psi_i$ is a solution. Evidently there are problems of convergence in both cases.

In this initial period only special integral equations were studied. The solution of general equations, that is for general kernels – possibly non symmetric or with discontinuities – was not attempted.

It was Du Bois-Reymond who draw attention to the necessity of developing a theory for these general equations in a paper in which he studied the solutions of the equation $\Delta u = 0$ (in 1888, see [8]). He introduced the name 'integral equation'. It is interesting to read Du Bois-Reymond (l.c. p. 228):

Ich schrieb diese Gleichungen (i.e. some integral equations in potential theory) nicht hin, als ob sie etwa das Problem lösten oder doch der Lösung näher führten, sie sollten nur ein Beispiel unter zahllosen sein, dafür, dass man bei Randproblemen der linearen partiellen Differentialgleichungen beständig vor dieselbe Gattung von Aufgaben gestellt wird, welche jedoch, wie es scheint, für die heutige Analysis im Allgemeinen unüberwindliche Schwierigkeiten darbieten. Ich meine die zweckmässig *Integralgleichungen* zu nennende Aufgaben, welche darin bestehen, dass die zu bestimmende

27

Function, ausser ihrem sonstigen Vorkommen, in ihnen unter bestimmten Integralen enthalten ist, welche ausser den Integralsvariabeln noch andere enthalten, die eigentlichen Argumente der Gleichungen, wie vorstehend α_0 und α_1. Eine einfache Form dieser Integralgleichungen, die viele besonderen Fälle umfasst, ist

$$\int_0^s ds f(s)\varphi(s, x) = \psi(x)\cdot f(x) + \chi(x)$$

zur Bestimmung von $f(x)$. φ, ψ, χ sind als bekannt zu denken. Aber es können wie im obigen Beispiele auch simultane Integralgleichungen für unsere Function vorgelegt sein. Eine einfache Integralgleichung für $\partial z/\partial N$ erhält man z.B., wenn man in

$$2\pi z = \int ds\left(z\frac{\partial R}{\partial N} - R\frac{\partial z}{\partial N}\right)$$

mit z an die Oberfläche geht. (...)

Doch werde ich noch viele solche Integralgleichungen, namentlich bei anderen partiellen Differentialgleichungen, aufzustellen Gelegenheit finden. Auf diese Aufgaben wurde ich zuerst durch meinen Freund *Adolph Fick*, den Würzburger Physiologen, aufmerksam gemacht, der, als wir vor mehr als 35 Jahren in Zürich zusammen arbeiteten, die Bemerkung machte, dass die Bestimmung von Dichtigkeiten, die von inneren Kräften der Substanz abhängen, vielfach darauf führt. Die Integralgleichungen sind wir seitdem in der Theorie der partiellen Differentialgleichungen so oft vorgekommen, dass ich überzeugt bin, die Fortschritte dieser Theorie seien an die der Behandlung der Integralgleichungen gebunden, über die aber so gut wie nichts bekannt ist.

In 1897 Poincaré [103] published a long paper, again on the method of Neumann in potential theory, in which he proved the convergence of the series in a more general situation. He introduced the parameter λ which since then has commonly been used:

$$\varphi(x) + \lambda\int_a^b K(x, y)\varphi(y)dy = \psi(x)$$

and studied, for the boundary value problem, the properties of the solution as a function of λ. Physical phenomena are the motivation of this parameter (vibrating membranes).[1]

Very important in the development of the theory of integral equations is the paper 'Sur une classe d'équations fonctionnelles' [37], published by Fredholm in 1903 in the *Acta Mathematica*. He studied the general equation

$$\varphi(x) + \int_0^1 K(x, y)\varphi(y)\mathrm{d}y = \psi(x),$$

where K and ψ are given functions, satisfying suitable conditions; the kernel K is integrable.

Fredholm referred to the work of Neumann and also to the publications on integral equations of Volterra in 1896. There is another source indeed for the theory of integral equations in the work of the Italian mathematicians Volterra and Pincherle at the end of the 19th century. In chapter III I will discuss the work of these mathematicians in more detail, because their work is related to another important concept in the development of functional analysis, namely the concept of a functional.

The above article from 1903 was preceded by three articles ([34], [35], [36]). In the note in the *Comptes Rendus* of 1902 Fredholm wrote:

On sait qu'un grand nombre de problèmes de la Physique mathématique conduisent à des équations fonctionnelles de la forme

$$\varphi(x_1, ..., x_n) + \int ... \int f(x_1, ..., x_n; y_1, ..., y_n)\varphi(y_1, ..., y_n)\mathrm{d}y_1...\mathrm{d}y_n =$$
$$= \psi(x_1, ..., x_n).$$

He introduced the symbolic notation

$$A_f\varphi(x) = \psi(x).$$

Just like Volterra, Fredholm was inspired by the analogy with the theory of systems of linear equations in algebra:

La théorie de l'équation (1) est un cas limite de la théorie des équations linéaires; aussi retrouve-t-on dans la théorie qui nous

---

[1] In this paper Poincaré introduced the denomination 'harmonic function' for the solutions of $\Delta u = 0$.

occupe tous les résultats de la théorie des déterminants. ([36], p. 219).

He introduced a

quantité D_K qui joue par rapport à l'équation fonctionnelle le même rôle que joue le déterminant par rapport à un système d'équations linéaires ([37], p. 367).

Introducing the notation

$$K\begin{pmatrix} x_1 \ x_2 \ \dots \ x_n \\ y_1 \ y_2 \ \dots \ y_n \end{pmatrix} = \begin{vmatrix} K(x_1, y_1) & K(x_1, y_2) & \dots & K(x_1, y_n) \\ K(x_2, y_1) & K(x_2, y_2) & \dots & K(x_2, y_n) \\ \cdot \ \cdot \ \cdot \ \cdot \ \cdot \ \cdot \ \cdot \ \cdot \ \cdot \\ K(x_n, y_1) & K(x_n, y_2) & \dots & K(x_n, y_n) \end{vmatrix}$$

Fredholm defined D_K by

$$D_K = \sum_{n=0}^{\infty} \frac{1}{n!} \int_0^1 \dots \int_0^1 K\begin{pmatrix} x_1 \ x_2 \ \dots \ x_n \\ x_1 \ x_2 \ \dots \ x_n \end{pmatrix} dx_1 dx_2 \dots dx_n.$$

D_K is the so called *Fredholm determinant.*

Using a theorem of Hadamard on determinants Fredholm showed that this series is convergent. He obtained the following results:

(i) *If $D_K \neq 0$, the equation has one and only one solution of φ.*

(ii) *The homogeneous equation, that is the equation in which $\psi = 0$, has a solution $\varphi \neq 0$ if and only if $D_K = 0$. If n is the order of the first 'minor' of D_K not equal to zero, then there are n linearly independent solutions.*

Furthermore Fredholm gave necessary and sufficient conditions for the inhomogeneous equation to have solutions in the case $D_K = 0$; they express a kind of orthogonality relations, showing analogy with well known conditions in algebra. The reader should compare these results with those in linear algebra.

Fredholm proved these results *directly*, using classical analysis.[1] I mean that he did not follow the way, as was done later by Hilbert, of considering an integral equation as the limit for $n \to \infty$ of a system of n linear equations, finding the solution of the integral equation as

1 For further information see Lalesco [80].

the limit of the solution of this finite system, the latter being solved with algebraical methods.[1]

In a short remark Fredholm mentioned the case in which the kernel K is replaced by a kernel λK, λ being a parameter, and he observed that in that case the determinant $D_{\lambda K}$ is an entire function of λ. But he did not study extensively the equations containing a parameter; this was done by Hilbert.

I mention the form in which the solution is obtained.

Let be given the equation

$$\varphi(x) + \lambda \int_0^1 K(x, y)\varphi(y)\mathrm{d}y = \psi(x),$$

in which K and ψ satisfy some conditions which I don't specify here. Define

$$D_K = \sum_{i=0}^{\infty} \frac{\lambda^n}{n!} \int_0^1 \dots \int_0^1 K\begin{pmatrix} x_1 \dots x_n \\ x_1 \dots x_n \end{pmatrix} \mathrm{d}x_1 \dots \mathrm{d}x_n$$

$$D_K(\xi, \eta) = K(\xi, \eta) + \sum_{n=1}^{\infty} \frac{\lambda^n}{n!} \int_0^1 \dots \int_0^1 K\begin{pmatrix} \xi\, x_1 \dots x_n \\ \eta\, x_1 \dots x_n \end{pmatrix} \mathrm{d}x_1 \dots \mathrm{d}x_n.$$

Then, if λ is not a solution of $D_K = 0$, the solution is determined by

$$\varphi(x) = \psi(x) + \lambda \int_0^1 S(x, y)\psi(y)\mathrm{d}y,$$

where

$$S(x, y) = -\frac{D_K(x, y)}{D_K}.$$

S is called the resolvent; like in algebra S is the quotient of two 'determinants'; S is a meromorphic function of λ. The roots of the equation $D_K = 0$ are called the *eigenvalues* and the corresponding solutions of the homogeneous equations *eigenfunctions*.

The work of Neumann, Poincaré and Fredholm was a motive for Hilbert to undertake the study of integral equations. His work, how-

---

1 The determinants I mentioned before occur already in the note [34] from 1900.

ever, contained considerably more than integral equations: in a later phase he placed them in the frame of functions of an infinite number of variables, of quadratic forms and of orthogonal transformations in an infinite number of variables. As to the connection between these subjects I remind the reader of the transformation on the axes of a quadratic form (conic section). This work became an important source for the theory of Hilbert space, of Banach spaces and of the theory of linear operators. It was published in six 'Mitteilungen' in the Göttinger Nachrichten in the years from 1904 till 1910. In 1912 these papers were added to a famous book *Grundzüge einer allgemeinen Theorie der linearen Integralgleichungen* [66]. The book contains not only Hilbert's theory on integral equations but also applications, for instance on the theory of partial differential equations and on complex function theory (Riemann's problem on the existence of complex functions with certain properties). There are also physical applications (kinetic theory of gases). Hilbert showed new ways and he introduced new notions which, in the hands of several mathematicians, among which E. Schmidt, F. Riesz, later developed into the notions of modern functional analysis. There are notions such as 'vollstetig' strong and weak convergence, a 'principle of choice' connected with weak compactness–be it that Hilbert did not use all of these denominations in this period. For more information about these notions see § 4.

With reference to the work on potential theory Hilbert gave the following motivation for his work on integral equations in the introduction to his book:

Die nähere Beschäftigung mit dem Gegenstande führte mich zu der Erkenntnis, daß der systematische Aufbau einer allgemeinen Theorie der linearen Integralgleichungen für die gesamte Analysis, insbesondere für die Theorie der bestimmten Integrale und die Theorie der Entwicklung willkürlicher Funktionen in unendliche Reihen, ferner für die Theorie der linearen Differentialgleichungen und der analytischen Funktionen sowie für die Potentialtheorie und Variationsrechnung von höchster Bedeutung ist. Ich beabsichtige in diesem Buche die Frage nach der Lösung der Integralgleichungen zu behandeln, vor allem aber den Zusammenhang und die allgemeinen Eigenschaften der Lösungen aufzusuchen, wobei ich meist die für meine Resultate wesentliche Voraussetzung mache, daß der Kern $K(s, t)$ der Integralgleichung eine *symmetrische* Funktion der Veränderlichen s, t ist. Insbesondere im vierten Kapitel gelange ich zu Formeln, die die Entwicklung einer willkürlichen Funktion nach gewissen ausgezeichneten Funktionen, die ich *Eigenfunktionen* nenne, liefern: es ist dies ein Resultat, in dem als spezielle Fälle die bekannten Entwicklungen nach trigonometrischen, Besselschen, nach Kugel-, Laméschen und Sturmschen Funktionen, sowie die Entwicklungen nach Funktionen mit mehreren Veränderlichen enthalten sind, wie sie zuerst Poincaré (a.a.O.) bei seinen Untersuchungen über gewisse Randwertaufgaben in der Potentialtheorie nachwies. *Meine Untersuchung wird zeigen, daß die Theorie der Entwicklung willkürlicher Funktionen durchaus nicht die Heranziehung von gewöhnlichen oder partiellen Differentialgleichungen erfordert, sondern daß die Integralgleichung es ist, die die notwendige Grundlage und den natürlichen Ausgangspunkt für eine Theorie der Reihenentwicklung bildet und daß eben jene erwähnten Entwicklungen nach Orthogonalfunktionen nur Spezialfälle eines allgemeinen Integralsatzes sind* – eines Satzes überdies, der als die direkte Erweiterung des bekannten algebraischen Satzes von der orthogonalen Transformation einer quadratischen Form in die Summe von Quadraten anzusehen ist. *Das merkwürdigste Resultat ist, daß die Entwickelbarkeit einer Funktion nach den zu einer Integralgleichung zweiter Art zugehörigen Eigenfunktionen als abhängig erscheint von der Lösbarkeit der entsprechenden Integralgleichung erster Art.*

Zugleich erhält dabei die Frage nach der Existenz der Eigenfunktionen eine neue und vollständigere Beantwortung. In dem besonderen Fall der Randwertaufgaben der Potentialtheorie hat bekanntlich die Existenz der Eigenfunktionen zuerst H. Weber auf Grund des Dirichlet-Thomsonschen Minimalprinzipes zu beweisen gesucht, und sodann hat Poincaré (a.a.O.) den Existenzbeweis mit Benutzung der von H. A. Schwarz ausgebildeten Methoden wirklich erbracht. *Durch Anwendung meiner Theoreme folgt nicht nur die Existenz der Eigenfunktionen im allgemeinsten Falle, sondern meine Theorie liefert zugleich in einfacher Form eine notwendige und hinreichende Bedingung für die Existenz unendlich vieler Eigenfunktionen.*

Die Methode, die ich in den folgenden Kapiteln I–VI anwende, besteht darin, daß ich von einem algebraischen Problem, nämlich dem Problem der orthogonalen Transformation einer quadratischen Form von n Variabeln in eine Quadratsumme, ausgehe und dann durch strenge Ausführung des Grenzüberganges für $n = \infty$ zur Lösung des zu behandelnden transzendenten Problemes gelange. Dieselben Theoreme über Integralgleichungen mit symmetrischem Kern werde ich in Kapitel XIV auf einem anderen Wege mittels der Methode der unendlichvielen Variabeln entwickeln.

In a footnote (*Grundzüge* p. 2) Hilbert refers to work of Volterra from 1897.

In this method there is some analogy with the procedure for solving an infinite system of linear equations with an infinite number of unknowns by considering such a system as the limit of a finite system; I described this method before. Now some details will be given in which Hilbert's notation (*Grundzüge*) is used.

Consider the integral equation

$$f(s) = \varphi(s) - \lambda \int_0^1 K(s, t)\varphi(t)\mathrm{d}t.$$

K is supposed to be symmetric and continuous in s and t. Hilbert introduced the following abbreviations:

$$K_{pq} = K\left(\frac{p}{n}, \frac{q}{n}\right), \quad (p, q = 1, 2, ..., n)$$

$$Kxy = K_{11}x_1 y_1 + K_{12}x_1 y_2 + K_{21}x_2 y_1 + ... + K_{nn}x_n y_n =$$
$$= \sum_{p, q} K_{pq}x_p y_q \quad (K_{pq} = K_{qp}),$$

$$\varphi_p = \varphi\left(\frac{p}{n}\right), \quad f_p = f\left(\frac{p}{n}\right), \quad (p = 1, 2, ..., n),$$

$$Kx_1 = K_{11}x_1 + K_{12}x_2 + ... + K_{1n}x_n,$$
$$Kx_2 = K_{21}x_1 + K_{22}x_2 + ... + K_{2n}x_n,$$
$$\cdot \quad \cdot \quad \cdot \quad \cdot \quad \cdot \quad \cdot \quad \cdot \quad \cdot \quad \cdot \quad \cdot \quad \cdot \quad \cdot \quad \cdot \quad \cdot \quad \cdot$$
$$Kx_n = K_{n1}x_1 + K_{n2}x_2 + ... + K_{nn}x_n,$$
$$[x, y] = x_1 y_1 + x_2 y_2 + ... + x_n y_n.$$

Then the following algebraic problem is the discretization of the problem of solving the integral equation:

$$f_1 = \varphi_1 - l(K_{11}\varphi_1 + \dots + K_{1n}\varphi_n),$$
$$f_2 = \varphi_2 - l(K_{21}\varphi_1 + \dots + K_{2n}\varphi_n),$$
$$\cdot \quad \cdot \quad \cdot \quad \cdot \quad \cdot \quad \cdot \quad \cdot \quad \cdot \quad \cdot \quad \cdot$$
$$f_n = \varphi_n - l(K_{n1}\varphi_1 + \dots + K_{nn}\varphi_n)$$

is a system of n linear equations in which $\varphi_1, \dots, \varphi_n$ are the unknowns, f_p, K_{pq} are given, l is a parameter.

The solution is determined by means of the classical method in the form of the quotient of determinants. There is some connection with the problem of orthogonal transformations of the quadratic form Kxx. For $n \to \infty$ Hilbert obtains in this way the formula that Fredholm proved earlier.

As an application Hilbert proves a theorem on the representation of a function–satisfying suitable conditions–in the form of a series

$$f = c_1\psi_1 + c_2\psi_2 + \dots$$

where the ψ_i are eigenfunctions of an integral equation and

$$c_i = \int_a^b f(s)\psi_i(s)\mathrm{d}s,$$

the so called *Fourier coefficients* with respect to the system (ψ_i).

The first chapters of Hilbert's book are however not the most important ones; in later chapters he treats questions which are much deeper and which were very important for further developments.

In these chapters Hilbert developed a theory of functions of an infinite number of variables. He gave a theory of orthogonal transformations of quadratic forms with an infinite number of variables and he applies his theory to integral equations, among which are equations with unsymmetrical kernel. In this work the importance of considering sequences (x_n) of real numbers x_n such that $\sum_{n=1}^{\infty} x_n^2$ is convergent, is observed.

The term 'Hilbert space' finds its justification in this contribution of Hilbert. In a certain sense the notion of convergence in the linear space l^2 of these sequences, the so called *strong convergence* defined by

$$\lim_{n \to \infty} \sum_{k=1}^{\infty} |x_k - x_k^{(n)}|^2 = 0,$$

goes back to Gauss, where it is connected with the method of least squares, but then there was evidently no anticipation of infinite dimensional spaces (Legendre also used the method of least squares; see [13]; see also [42]). Hilbert did not use in this work the terminology of infinite dimensional spaces either. The connection of the theory of integral equations and the theory of infinite systems of linear equations is shown by Hilbert by means of systems of orthogonal (orthonormal) functions, that are systems of functions φ_i, defined for instance in an interval $[a, b]$ (or on $\mathbb{R}$) and satisfying some regularity conditions, such that

$$\int_a^b \varphi_i \varphi_j \mathrm{d}x = \begin{cases} 1 & \text{if } i = j \\ 0 & \text{if } i \neq j \end{cases}.$$

Meanwhile the theory of the Lebesgue integral had been developed and Hilbert saw the importance of the measurable functions such that

$$\int |f(x)|^2 \mathrm{d}x$$

exists. Later this would lead to the space L^2 of such functions, which is isometric to the sequence space l^2, as is shown by means of the Fourier coefficients with regard to a basis of L^2. This is the key to Hilbert's theory of integral equations.

SOME DEFINITIONS

For the convenience of the reader I give some definitions of spaces which are important in the development that I described.

The space l^p $(p \geqq 1)$. The set of all sequences $x = (x_n)_{n \in N}$, $x_n \in \mathbb{R}$, such that

$$\sum_{n=1}^{\infty} |x_n|^p < \infty,$$

is a linear space over $\mathbb{R}$ which is denoted by l^p. A metric d is defined on l^p by means of the norm

$$\|x\| = (\sum_{n=1}^{\infty} |x_n|^p)^{1/p},$$

$$d(x, y) = \|x - y\|.$$

For $p = 2$ it is an infinite dimensional generalization of the euclidean space $\mathbb{R}^n$. The space l^2 is an example of the Hilbert space as it was defined axiomatically in a later stage of the development; this will be treated afterwards.

The space L^p ($p \geq 1$). The set of the equivalence classes of Lebesgue measurable real valued functions f, such that

$$\int_{\mathbb{R}} |f(x)|^p dx < \infty,$$

is a linear space over $\mathbb{R}$ which is denoted by L^p. A metric d is defined on L^p by means of the norm

$$\|f\| = (\int |f(x)|^p dx)^{1/p},$$

$$d(f, g) = \|f-g\|.$$

Measure theory is used in this definition, in particular sets of measure zero are important; an equivalence relation between functions is defined by means of such sets. L^2 is also an example of Hilbert space.

All these spaces are examples of a *Banach space*, a notion which was defined in the twenties when the axiomatic stage was attained. There are lots of examples of Banach spaces in analysis, but I confine myself to these examples. The development of the general notion of a linear space will be treated later;[1] its beginnings are already in the 19th century.

In a later section there will be again occasion to speak about this work of Hilbert in relation to the work of Schmidt, Helly and Riesz. It is worthwhile to mention already at this point that E. Schmidt—the mathematician I mentioned before; see his work on infinite systems of linear equations [113]—continued the work of Hilbert immediately in two papers entitled 'Zur Theorie der linearen und nichtlinearen Integralgleichungen', [112; I, II]. He gave the theory an entirely new basis. Just like in the case of the systems of linear equations his procedure is a *direct* one, that is to say he did not follow the way by which Hilbert (*Grundzüge* [66], chapters I–IV) obtained his results by considering the integral equation as the limit of a finite system of linear

---

1 For the definition of the concept of a linear space see chapter II.

equations. In the 'Einleitung' Schmidt writes (p. 435):

Nach Erledigung einiger Hilfssätze im ersten Kapitel der vorliegenden Arbeit finden im zweiten die Hilbertschen Sätze, unter Vermei-
dung des Grenzübergange aus dem Algebraischen, sehr einfache
Beweise.

His principal tools are systems of orthogonal functions and the well
known inequality of Bessel. I take from the paper of E. Hellinger
'Hilberts Arbeiten über Integralgleichungen und unendliche Gleichungssysteme' [61] the following quotation concerning this paper
of Schmidt (l.c. p. 104):

E. Schmidt hat in seiner Dissertation eine neue Begründung der
Hilbertschen Eigenwerttheorie gegeben, die an Übersichtlichkeit
und Kürze nicht übertroffen worden ist und den gröszten Einflusz
auf die weitere Entwicklung gewonnen hat. Er benutzt keine andere
mathematische Theorie, auch nicht die Hauptachsen-transforma-
tion der Algebra, und verwendet auch keine besondere Eigenschaft
des Integrals, sondern nur die Linearität der Integraloperationen
und damit unmittelbar zusammenhängende Eigenschaften; so ist
seine Methode unmittelbar auch auf mit anderen Prozessen gleicher
Eigenschaften gebildete Gleichungen, z.B. auf Gleichungen mit
endlichvielen Veränderlichen wie (1) oder mit unendlichvielen wie
in dem Problem van 8 anwendbar. Zum Überblick über die linearen
Lösungsscharen der Integralgleichungen werden stets orthogonale
Funktionensysteme als Bezugssysteme eingeführt (Schmidtsches
Orthogonalisierungsverfahren); als einziges spezifisches Konver-
genzhilfsmittel dienen die klassischen quadratischen Integralunglei-
chungen von H. A. Schwarz[1] und F. W. Bessel, alle Abschätzungen
erfolgen demnach durch die Integrale von Quadraten. Der Kern-
punkt der Methode ist der auf einer sukzessiver Approximation
beruhende direkte Existenzbeweis eines Eigenwertes und der zugehörigen Eigenfunktion, der dem klassischen Existenzbeweis von
H. A. Schwarz für den ersten Eigenwert der schwingenden Membran
nachgebildet ist. So erhält Schmidt alle Hilbertschen Resultate nebst

---

[1] The inequality of Schwarz is

$$(\int_a^b f(x)\varphi(x)\mathrm{d}x)^2 \leqq \int_a^b (f(x))^2\mathrm{d}x \cdot \int_a^b (\varphi(x))^2\mathrm{d}x.$$

der bereits erwähnten Erweiterung des Entwicklungssatzes, wobei ohne wesentliche Schwierigkeit auch unstetige aber *quadratisch integrierbare* Funktionen berücksichtigt werden können.

In [112; II] he gives very short proofs of the theorems of Fredholm. First he considers kernels of the form

$$K(s, t) = \sum_{i=1}^{m} \alpha_i(s)\beta_i(t),$$

and then he treats the general case by means of an approximation, using a theorem which we now call the theorem of Stone-Weierstrass.

In this work Schmidt is closer to the modern theory of linear operators–where it is the problem to find the inverse A^{-1} of an operator A–than Hilbert in his *Grundzüge*. But there is still no suggestion of an integral equation as a mapping from one function space into another, and in this respect it is classical analysis. There may be more indications of such an idea in Schmidt's paper of 1908 [113] on systems of linear equations which I mentioned before.

Hellinger's paper [61] gives an excellent survey of Hilbert's work on the theory of integral equations. In view of my purpose of studying the development of functional analysis–in which the integral equations were only a step, but an important step–I give no more details on the theory of these equations. I will come back to some of the notions which Hilbert introduced because they are useful for giving a picture of the development. With respect to Hellinger's paper [61] a few comments must be made.

On p. 107 Hellinger writes:

Er (Hilbert) führt folgende Begriffe ein (Grundz. Kap. XI, S. 147 ff., Kap. XII, S. 164 f., Kap. XIII, S. 174 ff.): Jedes Wertsystem x_p, $p = 1, 2, \ldots$ abzählbar unendlichvieler reeller Veränderlicher mit konvergenter Quadratsumme $\sum x_p^2$ wird als ein Punkt x eines unendlichdimensionalen Raumes H, der seither als *Hilbertschen Raum* bezeichnet wird, angesehen.

This is not quite correct. Perhaps Hellinger had the intention to describe in this sentence the deeper grounds by which Hilbert was led in establishing his theory, but nowhere in the *Grundzüge* the notion of an infinite dimensional space is introduced, not in the first chapters

39

in which Hilbert treats the integral equations by means of a limit procedure, neither in the last chapters where Hilbert deals with functions of an infinite number of variables. Here Hilbert introduced the infinitely many variables x_i, $i = 1, 2, \ldots$, and he says that he wants to restrict himself to the case that these variables satisfy the relation (*Grundzüge*, p. 126).

$$(x, x) = \sum x_i^2 \leq 1,$$

or more generally $\sum x_i^2 < \infty$ and he gives the formula

$$(u, v)^2 \leq (u, u)(v, v).$$

Hilbert then studied the theory of functions of these variables. But nowhere one finds any indication that thus he was doing analysis in a linear space. There is no indication that the domain of the variables is a linear space. Studying functions of infinitely many variables is not the same thing as studying geometry and analysis in an infinite dimensional space. Blumenthal writes in Hilbert's 'Lebensgeschichte' (*Gesammelte Abhandlungen* III, p. 412):

> Die Variabilität der abzählbar unendlich vielen Veränderlichen wird durch eine obere Grenze für die Summe ihrer Quadrate eingeschränkt (der später sogenannte 'Hilbertsche Raum').

The problem arises whether Hilbert was acquainted in this period (that is 1904–1910 or at least in the first years of this period) with the idea of an infinite dimensional linear space. There is all the more reason to pose this question because infinite dimensional linear spaces were already introduced at the end of the 19th century by the Italian school (Volterra, Pincherle, Peano), by Peano even in 1888. The only reference of Hilbert in the *Grundzüge* to the work of the Italian mathematicians is on page 2, where he refers to the work of Volterra on integral equations (the so called Volterra integral equations). But Volterra was not primarily interested in *linear* operators; he studied functionals ('fonctions de ligne'); Pincherle studied distributive operators. Did Hilbert know the work of the Italian school?

With respect to the introduction of function spaces in analysis it seems that Schmidt was closer to the modern point of view than Hilbert; in his paper of 1908 [113] he introduced 'lineare Gebilde', also without reference to the Italians. Some years later Riesz used the words 'totalité de fonctions' or 'Klasse', and in 1913 'espace', neither with reference to the Italian school. I already mentioned this in the section on infinite systems of linear equations.

There is another approach to the theory of integral equations namely from the side of the Italian mathematicians I already mentioned. Their work even preceded the results of Fredholm, Hilbert and Schmidt. Especially Volterra introduced a type of integral equations which are called *Volterra integral equations*. In a later chapter I will have occasion to speak about the Italian mathematicians where I shall be concerned with the axiomatic introduction of the notion of a linear space. But on this place I must give a short introduction because on the one hand Volterra integral equations are connected with linear functionals ('fonctions de ligne') and on the other hand linear functionals are in relation to the so called 'Moment problem', a problem that was important in the development of the ideas of Helly and Riesz, whose work was of fundamental importance to the introduction of the axiomatically defined normed linear spaces.

This source of the integral equations is much older. The origin is the so called calculus of variations, a part of mathematics that dates from the days of Euler and Lagrange. In this theory one is concerned with minimizing or maximizing a certain expression I, for instance an integral, which depends on a function $y = y(x)$ (or in terms of geometry: a curve) considered as variable and where f is a given function:

$$I = \int_a^b f(x, y, y') \mathrm{d}x.$$

One determines the so called variation δI of I and the condition $\delta I = 0$ leads in the well known way to a differential equation (equations of Euler-Lagrange). The classical examples from physics are well known. For example the problem of the Brachistochrone: a point-mass m moves from a point P_1 to a point P_2 under the influence of gravity; one asks to determine the curve between P_1 and P_2 such that the time if m moves on this curve is shortest. There are various isoperimetrical problems. And in mechanics the way of treating problems by means of variational principles (the equations of Hamilton).

About 1887 the Italian mathematician Volterra begins his publications on this subject. He observed that it is useful in analysis to study functions which depend on another function, that is to say on all the values of another function, well to be distinguished from the ordinary composed functions. Volterra published his results in several notes in 1887 [120]. In these notes Volterra deals with *funzioni dipendenti da altre funzioni*. In two further notes [121] he called them *funzioni dipendenti da linee*. The introduction of these mathematical objects is motivated by mechanics and physics. It is worthwhile reading the introductions of two books of Volterra, both published in 1913 and treating this subject [126], [127]. I took the following quotation from [127] p. 3:

Les fonctions qui dépendent d'autres fonctions ont une représentation du même type:[1] Par exemple envisageons toutes les lignes continues qu'on peut tracer dans un domaine à deux dimensions; en faisant correspondre à chaque ligne une valeur d'une variable, nous définirons *une fonction d'une ligne dans le domaine donné*. On peut avoir ainsi une représentation géométrique des fonctions qui dépendent de toutes les valeurs d'une fonction d'une variable. D'autre part, la Physique mathématique amène aussi à des notions concrètes sur le même sujet.

Imaginons un pôle magnétique fixe attirant un point magnétique *m*: la composante, par rapport à un axe, de l'attraction est *fonction du point m*; si *m* se trouve en présence d'un circuit électrique, on a une composante de l'attraction due au circuit, *fonction du point m*; mais si, *m* restant fixe, le circuit varie, la composante magnétique est *fonction de la ligne qui constitue le circuit*.

I quote some examples from [127] p. 8:

L'expression

$$I = \int_a^b \varphi(x)\mathrm{d}x,$$

est fonction de toutes les valeurs que $\varphi(x)$ prend dans l'intervalle (a, b). De même

---

1 Volterra makes a comparison with the 'ordinary' functions or 'functions of a point'. Fonction d'une ligne = function of a curve (line).

$$I = \int_a^b F(x)\varphi(x)\mathrm{d}x,$$

où la forme de $F(x)$ demeure constante. C'est la *forme de $\varphi(x)$* qui règle la variabilité de I. En d'autres termes, I dépend des ordonnées en nombre infini de la ligne

$$y = \varphi(x),$$

I dépend d'un nombre infini de variables.

At first (1887) Volterra introduced the notation

$$F|[\varphi(x)]|,_a^b$$

later he used (see [129], footnote p. 5):

$$F[\varphi(x)].\,_a^b$$

It is interesting to read what Fréchet says (1928) about the role of the calculus of variations in the development of functional analysis in the introduction to his book *Les espaces abstraits* ([32], p. 4):

La méthode classique, due à Lagrange, consiste non pas à traiter les fonctions de lignes de façon analogue aux fonctions de nombres, mais à passer par l'intermédiaire de fonctions qui, calculées sur une famille de lignes à un paramètre, se trouvent n'être que *des fonctions de ce paramètre numérique.* C'est aussi la méthode qui a été adoptée par M. Volterra pour les fonctions (de lignes) plus générale qu'étudie l'Analyse fonctionnelle. Nous croyons qu'on atteindra mieux le fond des choses et qu'on évitera des difficultés en abandonnant l'intermédiaire du paramètre et en traitant directement la ligne comme une variable absolument indépendante. Hâtons-nous d'ajouter qu'il n'en résultera aucune différence au point de vue formel et que les équations à resoudre seront les mêmes. Mais les théorèmes d'existence et la discussion en seront à notre avis facilités et éclairés. De plus, la théorie elle-même se trouvera rattachée à un ordre d'idées plus général. (...)

Une remarque analogue s'applique à l'utilisation de la théorie des *fonctions à une infinité de variables* dans la solution des problèmes de l'Analyse fonctionnelle.

La méthode qui a été employée avec succès par MM. Volterra et Hilbert consiste à remarquer qu'une fonction (continue, par exemple) peut être déterminée par la connaissance d'une infinité dénombrable de paramètres. Et alors on traitera d'abord le problème comme si l'on n'avait à faire intervenir qu'un nombre fini de paramètres, puis on passera à la limite.

Nous croyons que cette méthode a joué un rôle utile en secondant l'intuition, mais qu'elle a fini son temps. C'est un artifice inutile de substituer à la fonction une suite infinie de nombres qui, d'ailleurs, peut être choisie de plusieurs façons.[1] On le voit bien, par exemple, dans la théorie des équations intégrales où les solutions de Fredholm et de Schmidt sont beaucoup plus simples et plus élégantes que celle de Hilbert, ce qui n'enlève pas à ce dernier le mérite essentiel d'avoir obtenu par sa méthode un grand nombre de résultats nouveaux.

En résumé, la méthode la plus féconde en Analyse fonctionnelle nous paraît être celle qui consiste à traiter l'élément dont dépend la fonctionnelle, directement comme une variable et sous la forme même où il se présente naturellement. Cela permettra d'utiliser immédiatement un grand nombre de propositions actuellement établies en Analyse générale. Et cela évitera d'introduire des éléments parasites (paramètres, coefficients) susceptibles d'amener des complications inutiles et étrangères au fond du problème.

In 1903 Hadamard introduced the term 'fonctionnelle' or functional –thus essentially coining the present name of the subject–instead of 'fonction de ligne'. I will return to this in § 5 of this chapter.

For the class of functions on which a functional is defined Volterra used in his book *Théorie générale des fonctionnelles* [129], p. 8, the term 'champ fonctionnel':

L'étude des champs fonctionnels, c'est à dire des ensembles dont

---

1 Fréchet added the following footnote: 'Notre critique doit être entendue dans le même sens et avec les mêmes limitations que celle qui s'adresse à un emploi abusif des coordonnées dans les questions qui relèvent de la géométrie pure.' Compare the remarks on p. 92 on coordinate-free methods and my quotations of Poincaré there.

les éléments sont des fonctions, est évidemment d'un intérêt primordial pour une compréhension du concept de fonctionnelle.

Pour une fonctionnelle $F[\overset{b}{\underset{a}{x(t)}}]$, le champ fonctionnel le plus vaste est celui de toutes les fonctions possibles $x(t)$ définies dans l'intervalle (a, b): il constitue en somme un 'espace' à une infinité non dénombrable de dimensions (le nombre de ses dimensions a la puissance du continu en ce que chaque élément de cet ensemble est défini par un ensemble de valeurs de x correspondantes à toutes les valeurs de t entre a et b).

Another 'champ fonctionnel' is the set of all analytic functions, or all continuous functions, defined on an interval (a, b). But when this was written (the book dates from 1936), the notions of a function space and of infinite dimensional spaces were already reasonably well known among mathematicians (Banach's book *Théorie des opérations linéaires* dates from 1932).

Now let us return to Volterra's work which led him at the end of the 19th century to study integral equations. Volterra started to develop the analysis of functionals. He defined the concept of continuity for functionals. For defining continuity it must be agreed when we shall say that the functions (curves) φ_n converge to the function φ. This may be uniform convergence but sometimes it is possible to define a hierarchy of types of convergence. This anticipates later topological considerations and, in particular, the introduction of different systems of neighborhoods of a point.[1] Volterra did not yet define his concepts in terms of topology.

Volterra studied differentiation of–not necessarily linear–functionals and he defined the variation of a functional F; I give the result without an explanation (the reader may compare the analogy with the theory of 'ordinary' functions of several variables):

$$\delta F = \int_a^b F'|[f(x), \xi]|\delta f(\xi)\mathrm{d}\xi.$$

[1] Hadamard ([52], p. 49) remarks that the concept of types of convergence goes back to Weierstrass. One can require, for instance, the convergence of the values of the functions and also the convergence of the derivatives up to some order.

45

In 1887 this led Volterra to the introduction of a type of integral equations which were afterwards called *Volterra integral equations*. There are two types:

$$\varphi(x) = u(x) + \lambda \int_0^x K(x, \xi)u(\xi)\mathrm{d}\xi,$$

$$\varphi(x) = \lambda \int_0^x K(x, \xi)u(\xi)\mathrm{d}\xi.$$

The difference with the Fredholm equations is that now the upper limit in the integral is variable.

In several notes Volterra succeeds in solving equations of these types; I mention his paper 'Sulla inversione degli integrali definiti' [123]; for a complete bibliography I refer the reader to [129]. Note that Volterra refers to the work of Abel on integral equations.

The leading idea of Volterra for the solution of integral equations was the analogy with the algebraic problem of solving a system of linear equations. It seems that he was the first who initiated this approach. I quote from [127], p. 33:[1]

M. Volterra étant guidé par les idées que nous venons d'exposer, qui découlaient de sa *théorie des fonctions des lignes*, a été le premier à considérer les équations intégrales comme le cas limite d'un nombre infini d'équations algébriques et à donner la solution générale de l'équation de son type en partant de cette conception.

In his paper from 1896 [123] Volterra deals with the equations

$$f(y) - f(\alpha) = \int_\alpha^y \varphi(x)H(x, y)\mathrm{d}x.$$

and he remarks that the solution φ can be considered as the limiting case of the solution of a system of the following kind:

$$b_1 = a_{11}x_1$$
$$b_2 = a_{12}x_1 + a_{22}x_2$$
$$b_3 = a_{13}x_1 + a_{23}x_2 + a_{33}x_3$$
$$\cdot \;\cdot\;\cdot\;\cdot\;\cdot\;\cdot\;\cdot\;\cdot\;\cdot\;\cdot\;\cdot\;\cdot\;\cdot\;\cdot$$
$$b_n = a_{1n}x_1 + a_{2n}x_2 + a_{3n}x_3 + \ldots + a_{nn}x_n,$$

1 The book contains *Leçons professées à la faculté des Sciences de Rome en 1910*, which were published by M. Tomassetti and F.-S. Zarlatti. Hilbert is quoted in this book.

in which the a_{is} correspond to $H(x, y)$.

In [127], p. 31 and 32, where the equation (marked by (3)):

$$\mu\varphi(x) = \lambda u(x) + \lambda \int_a^b K(x, \xi)u(\xi)d\xi.$$

is considered, Volterra states:

L'idée *toute naturelle* qui est suggérée par la considération des fonctions des lignes est de considérer l'équation linéaire (3) comme cas limite d'une expression algébrique, obtenue en partageant l'intervalle (a, b) en n intervalles $h_1, \ldots, h_n$,

$$\varphi(x) = u(x) + \lim_{n=\infty} \sum_{i=1}^{n} K(x, \xi_i)u(\xi_i)h_i,$$

la résolution de l'équation intégrale (3) *est ainsi ramenée à l'étude d'une équation algébrique linéaire à une infinité d'inconnues.*

Avant de passer à la limite pour $n = \infty$ et en donnant à x les valeurs $\xi_1, \xi_2, \ldots, \xi_n$, on peut écrire, au lieu de l'équation intégrale (3), le *système algébrique*

$$\varphi(\xi_s) = u(\xi_s) + \sum_{i=1}^{n} K(s_i, \xi_i)u(\xi_i)h_i \quad (s = 1, 2, \ldots, n).$$

En résolvant ce système on trouve les valeurs de

$$u(\xi_1), u(\xi_2), \ldots, u(\xi_n)$$

correspondantes à l'équation (3) *et la limite de la solution, si elle existe, pour $n = \infty$ sera solution de l'équation intégrale proposée.*

Fredholm, referring to the papers of Volterra from 1896, must have been led by the same idea; it is the idea by which the determinants I mentioned on page 30 came into the theory.

In the case of the equations of the Volterra type (with variable upper limit in the integrals) the determinant D_K equals 1 and the solutions are, as a consequence of this fact, holomorphic functions of λ.

All this should be seen in the frame of the general theory of functionals. It was an important extension of classical analysis and it was a step towards the study of the general properties of classes of functions, of function spaces, i.e. sets of functions on which a certain structure is

47

defined, for instance a linear structure or a topological structure, and towards the introduction of abstract spaces and structures in analysis. The set of linear functionals appears there as a *dual space*.

I will come back to this development in a later chapter where I shall discuss the work of Peano, Pincherle, both mathematicians belonging to what I called the Italian school. Furthermore I must speak about the 'analyse générale' of Fréchet, which is a very general kind of abstract analysis.

§ 3 *The problem of moments*

The origin of the so called problem of moments is in the theory of probability. In a modern version it is the following problem.

Let be given a sequence (c_n) of real numbers. One asks to determine an increasing function ξ, defined on the interval $a \leq x \leq b$ such that

$$\int_a^b x^n d\xi(x) = c_n$$

for $n = 1, 2, \ldots$[1]

It is evident that this problem has no solution for arbitrary values of c_n. Indeed, for having a solution it is necessary that $c_n \geq 0$ for all n; but this condition may not be sufficient and in general it is not. So, a better version of the problem is to find necessary and sufficient conditions for the existence of a solution.

This problem was solved in the course of time: one knows necessary and sufficient conditions. The problem has several connections with other parts of analysis: continued fractions, orthogonal systems of functions, quadratic forms, infinite matrices, quasi-analytic functions. But it is beyond the scope of this book to treat these connections. I refer the reader who wants to study them to the book of Shohat and Tamarkin [114]. The connection of this problem with the theory of probability was pointed out by the Russian mathematician Tchebycheff, who started studying this subject about 1855. He considered

---

[1] This is the *Hausdorff moment problem*. There are some other problems in which the interval $[a, b]$ is not finite (Stieltjes problem and Hamburger problem). The respective theories diverge.

the integrals

$$\int\limits_{-\infty}^{+\infty} f(x)x^n \mathrm{d}x, \quad n = 0, 1, 2, \ldots,$$

in which f is non-negative. These are the moments of the distribution on $(-\infty, +\infty)$ determined by the function (density) f. Note that he did not yet define the moments in terms of a Stieltjes integral, as was done in the above formulation of the problem of moments. That type of integrals, later called Stieltjes integrals, was introduced by Stieltjes in 1894 in a famous paper on continued fractions [119]. He already solved the problem of moments for the case of integrals, taken from 0 to ∞. Tchebycheff's interest in this problem was motivated by some notions in probability theory, in particular by the notion of normal distribution. Because these distributions are defined by means of the function $x \to e^{-x^2}$, he asked whether from

$$\int\limits_{-\infty}^{+\infty} f(x)x^n \mathrm{d}x = \int\limits_{-\infty}^{+\infty} e^{-x^2}x^n \mathrm{d}x$$

for $n = 0, 1, 2, \ldots$, one may conclude that $f(x) = e^{-x^2}$. That is: is the normal distribution characterized by the set of its moments?

However, it is not at this point that the problem of moments is connected with the development of functional analysis. This connection is to be found in the problem of the existence of continuous linear functionals on linear function spaces and, more generally, in the theory of duality in Banach spaces. This development has gone a long way which ultimately led to the theorem of Hahn-Banach (1927).

Earlier I spoke about functionals ('fonctions de ligne'). The notion of a *linear* functional is evidently connected with the notion of a *linear space*.[1] For the sake of convenience I give some definitions.

Let E be a linear space over the field of the reals $\mathbb{R}$. *A linear functional*

---

1 For the definition of the concept of a linear space see chapter II. Some authors speak of a vector space instead of a linear space; I use both terms. Instead of 'linear functional' the term 'linear form' or 'linear function' is used. There is indeed no reason to distinguish between functions and functionals. Both are examples of the general concept of a mapping from one space into another. The word functional is used on historical grounds.

T is a mapping from E into $\mathbb{R}$ satisfying

$$T(ax) = aT(x), \quad a \in \mathbb{R}, \ x \in E,$$

$$T(x+y) = T(x) + T(y), \quad x, y \in E.$$

When E is a topological linear space–for instance a normed space–it makes sense to speak of *continuous* linear functionals; continuity is defined in the well known way.

A linear space E over $\mathbb{R}$ is called a normed space over $\mathbb{R}$ when there is defined a mapping $\|\cdot\|$ from E into $\mathbb{R}$ satisfying

$$\|x\| = 0 \Leftrightarrow x = 0,$$

$$\|ax\| = |a| \cdot \|x\|, \quad a \in \mathbb{R}, \ x \in E,$$

$$\|x+y\| \leqq \|x\| + \|y\|.$$

A metric d on a normed space is defined by

$$d(x, y) = \|x-y\|.$$

The spaces l^p and L^p which I defined before are examples of normed spaces.

One proves that continuity of the linear functional T in normed spaces is equivalent with boundedness of T, i.e. the existence of a constant $M \geqq 0$ such that

$$|T(x)| \leqq M \cdot \|x\|$$

for all $x \in E$.

Let E and F be normed spaces over $\mathbb{R}$. A linear operator T is a mapping from E into F such that

$$T(ax) = aT(x), \quad a \in \mathbb{R}, \ x \in E$$

$$T(x+y) = T(x) + T(y).$$

Continuity of T is equivalent with boundedness, i.e.

$$\|T(x)\| \leqq M\|x\|,$$

where in the left member $\|\cdot\|$ is the norm in F and $\|\cdot\|$ in the right member is the norm in E.

I must warn the reader that these are the modern definitions, formulated in the framework of an axiomatic theory of normed spaces. It differs totally from the situation in the first years of the development towards modern functional analysis. The leading mathematicians on this subject were in the first years Riesz and Helly.

The introduction of functionals and distributive linear operators goes back to the 19th century when they were studied by Pincherle and Volterra. However, for the connection between the problem of moments and linear functionals I must turn first to Hadamard who, in his paper in the *Comptes Rendus* of 1903 (Hadamard, *Oeuvres* I, p. 405) remarks:

Les opérations fonctionnelles linéaires, c'est à dire les lois suivant lesquelles on peut, à toute fonction $f(x)$ définie dans un intervalle $a < x < b$ faire correspondre un nombre U, de telle façon qu'on ait (quels que soient les nombres c_1, c_2 et les fonctions f_1, f_2)

$$U(c_1 f_1 + c_2 f_2) = c_1 U(f_1) + c_2 U(f_2)$$

ont été étudiées principalement par MM. Volterra, Pincherle, Bourlet (for Bourlet see [13]).

Note that Hadamard does not state that, in order that this definition makes sense, it is necessary that if f_1 and f_2 belong to the family of functions under consideration $c_1 f_1 + c_2 f_2$ must belong to the family, in other words that the set of functions form a linear space. Hadamard did not use the terminology of spaces. This is the more remarkable because the notion of an infinite dimensional linear space was already known at the time that Hadamard wrote this paper: as we will see later Peano gave a definition in 1888 and Pincherle wrote a book on the subject in 1901. On the other hand Hadamard did refer to Pincherle.

In this paper [51] Hadamard studied the form of the linear functionals U on the family (space) of the continuous functions defined on $[a, b] = I$. I denote this space by $C(I)$. It is customary to define the norm by $\|f\| = \sup_{x \in I} |f(x)|$. He found the representation (l.c. p. 407):

$$U[f(x)] = \lim_{\mu \to \infty} \int_a^b f(x) \Phi(x, \mu) dx,$$

in which the Φ are continuous functions, which depend on U but not on f.

Riesz continued the work of Hadamard in a paper from 1911 [108]. In this paper functions of bounded variation play an important role in connection with the theory of the Stieltjes integral.[1] He poses the following problem:

Etant donnée une fonction $A(x)$, il faut reconnaître s'il existe ou non une fonction à variation bornée $\alpha(x)$ dont elle est l'intégrale indéfinie.

This leads him to consider the set of all continuous functions defined on the interval $[a, b]$; he speaks about the 'totalité des fonctions continues $f(x)$'. He states the result of Hadamard in the form: there are continuous functions a_n such that

$$A[f(x)] = \lim_{n \to \infty} \int_a^b a_n(x)f(x)\mathrm{d}x.$$

In this paper Riesz proved the following result, which is an important improvement of the theory of Hadamard.[2] Defining 'Toute opération distributive et continue sera dite linéaire', he proves ([108], p. 43):

Etant donnée l'opération linéaire $A f(x)$, on peut déterminer la fonction à variation bornée $\alpha(x)$ telle que pour toute fonction continue $f(x)$ on ait

$$A[f(x)] = \int_a^b f(x)\mathrm{d}\alpha(x).$$

Then Riesz considers the system

$$\int_a^b f_k(x)\mathrm{d}\alpha(x) = c_k \quad (k = 1, 2, \ldots),$$

in which the continuous functions f_k are given and the function α, which is supposed to be of bounded variation, must be determined. He finds necessary and sufficient conditions of the form I mentioned on p. 15.

Now it is clear that the problem of moments can be formulated as

---

1 I remind the reader that any function of bounded variation is the difference of two increasing functions.

2 Riesz obtained this result in 1909; see [106]; the paper [108] is dated june 1910 and refers to [106]. Compare [30]. This theorem is important in the development of the concept of integral (see [20]).

the problem of the existence of a linear functional on a linear space which takes given values for a given sequence of functions (elements of the space). Taking for the given elements the functions $x \to x^n$ one obtains the problem of moments. The problem is whether a linear functional on a space E is already determined by its values on a subset of E. On $C(I)$, for instance, it is a simple consequence of the theorem of Stone-Weierstrass that a continuous linear functional is defined by the values it takes for the functions $x \to x^n$. I shall discuss the theorem of Stone-Weierstrass in chapter IV.

Riesz seems to have derived some inspiration from the problem of moments as may be concluded from the following passage (l.c. [108], p. 54):

Cherchons maintenant les conditions de ce que le système (11) admette de solution monotone, cette recherche nous étant suggérée par plusieurs problèmes bien connus de l'analyse qui y sont liés; il suffira de citer le *problème des moments* de Stieltjes, les recherches fondamentales de M. Hilbert concernant les *formes quadratiques d'une infinité de variables* et les beaux et importants travaux de M. Carathéodory qui se rattachent au théorème de Picard-Landau.

In proving his representation theorem Riesz used the following lemma ([108], p. 41):

$A[f(x)]$ désignant une opération linéaire quelconque, on y peut attacher un nombre M_A tel que pour toute fonction continue $f(x)$

$$|A[f(x)]| \leqq M_A \times \text{maximum de } |f(x)|.$$

Evidently this is the relation between continuity and boundedness of the linear functionals (see p. 50).

Observe that there is a relation between the problem of moments and the theory of infinite systems of linear equations which I discussed in § 1. Indeed, in any sequence space E such that the linear functionals are of the form $\sum_{n=1}^{\infty} a_n x_n$, $(x_n) \in E$, $a_n \in \mathbb{R}$, the problem of moments reduces to the problem of solving a system

$$\sum_{i=1}^{\infty} a_{ik} x_i = c_k, \quad k = 1, 2, \ldots.$$

The mathematician Helly, whose work I will treat afterwards, occupied himself with this problem.

There is another paper of Riesz, written a year earlier (1910), which is interesting [107]. Referring to Stieltjes he studies in this paper the system

$$\int_a^b f_i(x)\xi(x)\mathrm{d}x = c_i, \quad i = 1, 2, \ldots,$$

in which f_i, c_i are given and ξ has to be determined. He refers to the preceding work of Hilbert, Schmidt and Fischer, who considered functions whose square is integrable. Here Riesz studies the problem for a more general class of functions (he did not use the word 'space'), namely for the class ('Klasse') of functions f such that $|f|^p$ is integrable in the sense of Lebesgue. He denotes this class by $[L^p]$ ($p > 1$). Although Riesz does not formulate the theory with equivalence classes of functions, he remarks:

> wir dürfen in den folgenden Entwicklungen zwei Funktionen, die höchstens mit Ausnahme einer Nullmenge übereinstimmen d.h. deren Differenz eine Nullfunktion ist, als identisch betrachten.

Riesz states the linearity of this class ([107], p. 459):

> Für das weitere ist es noch wichtig, festzustellen, dasz *die Klasse* $[L^p]$ *alle linearen Verknüpfungen je einer endlichen Anzahl in ihr enthaltener Funktionen ebenfalls enthält.* Dies folgt aus Ungleichung (8).[1]

Riesz requires $f_i \in [L^{p/(p-1)}]$ and he wants to have $\xi \in [L^p]$. Here we already see (in 1910) the connection between the spaces (in modern notation) L^p and L^q with $1/p + 1/q = 1$. I mention the following passage in Riesz's paper ([107], p. 452):

> Die Untersuchung dieser Funktionenklassen (i.e. L^p) wird auf die wirklichen und scheinbaren Vorteile des Exponenten $p = 2$ ein ganz besonderes Licht werfen; und man kann auch behaupten, dasz sie für eine axiomatische Untersuchung der Funktionenräume brauchbares Material liefert.

The conclusion seems justified that Riesz, when writing this (that is in 1910), had an idea of something like an *axiomatic theory*. Some years

---

[1] This is the inequality

$$\left[\int \left|\sum_{i=1}^n c_i f_i(x)\right|^p \mathrm{d}x\right]^{1/p} \leqq \sum_{i=1}^n |c_i| \left[\int |f_i(x)|^p \mathrm{d}x\right]^{1/p}.$$

earlier Riesz already had the idea of a geometry in infinite dimensional spaces, as may be seen from a paper from 1907: 'Sur une espèce de géométrie analytique des systèmes de fonctions sommables' [105] in which he observes that the purpose of his studies is 'Approfondir la méthode des coordonnées appliquée à l'étude des systèmes de fonctions sommables' (there are some earlier notes containing references to Fréchet).

The idea of function spaces and infinite dimensional spaces was in the mind of a number of mathematicians in those years. I mention the papers of Fréchet. For instance 'Sur les ensembles de fonctions et les opérations linéaires' ([30]; 1907). In this note Fréchet considers 'le champ (R) des fonctions sommables et de carrés sommables, définies par exemple dans l'intervalle $(0, 2\pi)$'; he states that he has determined the general form of the continuous linear functionals ('opérations linéaires') in this 'champ de fonctions'. Further Fréchet's paper 'Essai de géométrie analytique à une infinité de coordonnées', published in 1908 (see [32], p. 283).[1]

One would be inclined to say that this fact is not at all surprising because at the end of the 19th century Peano, Volterra, Pincherle already were explicitly operating with the notion of function spaces and even of linear spaces in an axiomatic setting (see chapter II). But although Riesz does refer to Hadamard's paper from 1903–where Hadamard refers to the Italian mathematicians–there is no reference in the work of Riesz to the papers of these Italian mathematicians. Was Riesz acquainted with their work?

However that may be, it seems that Riesz was in his work of those years nearer to modern functional analysis than Hilbert in his *Grundzüge*. There is no explicit indication in the *Grundzüge* of the idea of a family of functions characterized by certain properties of these functions. On the other hand, as I observed before, Hilbert refers to a paper of Volterra from 1897 [124], but not to the work of Peano and Pincherle–who published on linear spaces whereas Volterra was perhaps more interested in 'fonctions de ligne' at that time–nor to Hadamard (1903). Was Hilbert more interested in the application of the theory of integral equations on other parts of mathematics–which

---

1 The theory of abstract spaces is a subject of general analysis; see chapter III. There is an extensive bibliography in Fréchet [32].

are numerous in the *Grundzüge*–than in the abstract theory behind it? See for this [104], p. 105.

Finally I must deal with the work of the Austrian mathematician E. Helly. It seems somewhat forgotten nowadays that he published some important papers on 'lineare Funktionaloperationen' and systems of linear equations (1912, 1921) in which he came close to the theorem of Hahn-Banach, the famous theorem on the existence of continuous linear functionals. Helly is mainly known by some work on convex sets (see p. 69).

In his paper 'Über lineare Funktionaloperationen' from 1912 [64] Helly continued the work of Riesz (1911, [108]) on the representation of the linear functionals on the space of continuous functions. His aim is to give a new proof of the representation theorem of Riesz:

> Im folgendem soll zunächst für diesen zweiten Riesz'schen Satz ein Beweis gegeben werden, der die von Riesz selbst benützten etwas umständlichen und nicht ganz in der Sache liegenden Abschätzungsmethoden vermeidet und nur Hilfsmittel benützt, die durch die Theorie der linearen Funktionaloperationen selbst dargeboten werden. Ferner wird sich daraus ein Beweis des Hadamard'schen und ersten Riesz'schen Satzes ergeben. Und endlich soll eine Anwendung der gefundenen Resultate auf die sogenannte Integralgleichung erster Art gemacht werden. ([64], p. 266)

I quote some passages from this important paper.

> Eine eindeutige Funktionaloperation wird dadurch definiert, dasz jeder Funktion f eines gewissen Bereiches E eine bestimmte Zahl $U[f]$ zugeordnet wird. Die Funktionaloperation soll stetig genannt werden, wenn
>
> $$\lim_{i=\infty} U[f_i] = U[f] \tag{1}$$
>
> ist, wenn die Funktionen f_i die Funktion f zur Grenze haben, wobei die Art des Grenzüberganges noch näher festgelegt werden kann.
>
> Im folgenden soll, wenn nicht ausdrücklich etwas anderes verlangt wird, die Geltung der Beziehung (1) nur dann gefordert werden, wenn die Funktionen f_i gleichmäszig gegen f konvergieren.
>
> Eine stetige Funktionaloperation soll linear genannt werden,

wenn für je zwei Funktionen f und g des Bereiches E immer die Beziehung

$$U[f+g] = U[f] + U[g] \qquad (2)$$

besteht. Für das folgende soll der Bereich E aus allen in einem bestimmten Intervalle $a \leqq x \leqq b$ stetigen Funktionen bestehen.[1]

This second problem of Riesz is the following problem which I represent here in the form Helly gave it ([64], p. 271):

IV. Die notwendige und hinreichende Bedingung dafür, dasz eine lineare Funktionaloperation $U(f)$ existiert, deren Maximalzahl den Wert M nicht übersteigt, und für die

$$U(f_i) = c_i \quad (i = 1, 2, \ldots)$$

ist, besteht darin, dasz die Ungleichung

$$\left| \sum_1^n i\, \mu_i c_i \right| \leqq M\, \overline{\left| \sum_1^n i\, \mu_i f_i(x) \right|}$$

für alle Werte der Zahlen μ_i und jedes n erfüllt ist.

The 'Maximalzahl' needs some explanation. It is connected with the relation expressing the equivalence of continuity and boundedness (see p. 53). Helly defines the 'Maximalzahl' as the smallest number M for which

$$|U[f]| \leqq M\, \overline{|f(x)|},$$

denoting by

$$\overline{|f(x)|}$$

the maximum of the absolute value of $f(x)$ in the interval (a, b). This 'Maximalzahl' is what later will be called the *norm* $\|U\|$ of the functional U.

In the proof essential use is made of the following theorem ([64], p. 267):

I. Aus jeder unendlichen Menge linearer Funktionaloperationen,

---

1 Note that Helly used here the notation f, g for functions and not, as was common in those days, $f(x)$, $g(x)$. In his definition of a 'stetige Funktionaloperation' Helly does not specify the condition $U[\alpha f] = \alpha U[f]$, but in the proofs of his theorems he makes use of this relation.

deren Maximalzahlen unter einer endlichen Grenze M liegen, läszt sich eine Teilreihe $U_1[f]$, $U_2[f]$, $U_3[f]$ ... herausgreifen, die gegen eine lineare Operation konvergiert, deren Maximalzahl nicht gröszer als M ist.

This is also a theorem for which Helly refers to Riesz ([108], p. 49); Riesz called it a 'principe de choix'.[1] In modern mathematical language this property expresses the weak compactness of the unit sphere in the dual space. There are some more theorems in this remarkable paper of Helly in which he anticipates the modern (axiomatic) theory; for instance ([64], p. 268):

III. Wenn eine unendliche Reihe linearer Operationen $U_i(f)$ gegeben ist, so dasz $\lim_{i=\infty} U_i(f)$ für jede stetige Funktion $f(x)$ existiert und selbst wieder eine lineare Funktionaloperation ist, so liegen die Maximalzahlen sämtlicher Operationen $U_i(f)$ unter einer endlichen Grenze.

This is evidently a form of the theorem of Banach-Steinhaus (see chapter IV). Helly then proved the representation theorems of Hadamard and Riesz. All this shows the advance (in 1912) towards modern functional analysis, but the theory has not yet reached the modern axiomatic form.

§ 4 *Supplementary remarks on the work of Hilbert, Riesz and Helly*

In this section I will treat some details from the work of Hilbert, Riesz and Helly, in order to show some special features of the move towards modern functional analysis. Especially I will treat types of convergence.

Again I consider Hilbert's *Grundzüge* that I discussed before. But then I passed over some notions which were important in the development.

---

[1] In a footnote on page 57 in [109], Riesz refers for this 'principle' to Fréchet: 'Voir pour ce principe, son historique et son rôle en Analyse: Fréchet, Sur quelques points du calcul fonctionnel, Thèse, Paris, 1906 (impr. aussi dans les Rendiconti del Cir.mat. di Palermo, t. XXII, 1906), Chap. IV–VII.' It is not clear why Riesz called it a principle and not a theorem, all the more because Riesz gave a proof of the result. There are more 'principles' in mathematics whichs are theorems, for instance the Principle of Dirichlet.

Essential points in the *Grundzüge* are the functions of an infinite number of variables and the quadratic forms in such a system of variables, that is to say the forms

$$\sum_{i,k=1}^{\infty} a_{ik}x_i x_k$$

or, in the complex case

$$\sum_{i,k=1}^{\infty} a_{ik}x_i \bar{x}_k, \quad a_{ik} = \bar{a}_{ki}.$$

The theory of such quadratic forms, in relation to the infinite systems of linear equations and the integral equations, is an important subject in the *Grundzüge*, especially in the later parts.

The connection between the integral equations and the linear systems is established by means of complete systems of orthogonal functions. Here the isometry of l^2 and L^2 is fundamental. This theorem is known under the name of the *theorem of Fischer-Riesz*: *to every element* $(a_k) \in l^2$ *there is a uniquely determined function (better: class of functions)* $f \in L^2$, *corresponding to* (a_k), *such that the* a_k *are the Fourier coefficients of this f with respect to an orthogonal system* (φ_i) *and conversely.*

Here the convergence of the series $\sum a_i \varphi_i$ is to be taken in the sense of norm convergence in L^2.

Hilbert gave a theory of eigenvalues of these quadratic forms and a decomposition in squares; this means that he developed a spectral theory for the infinite quadratic forms.

The most important notion Hilbert used for this, is the notion *vollstetig*.[1] Because of its importance I give the definition; see *Grundzüge*, p. 147, 175.

Wir nennen eine Funktion $F(x_1, x_2, ...)$ der unendlich vielen Variablen $x_1, x_2, ...$ für ein bestimmtes Wertsystem desselben *vollstetig*, wenn die Werte von $F(x_1 + \varepsilon_1, x_2 + \varepsilon_2, ...)$ gegen den Wert $F(x_1, x_2, ...)$ konvergieren, wie man auch immer $\varepsilon_1, \varepsilon_2, ...$ für sich zu Null werden läszt, d.h. wenn

$$\underset{\varepsilon_1=0, \varepsilon_2=0, \cdots}{L} \quad F(x_1 + \varepsilon_1, x_2 + \varepsilon_2, ...) = F(x_1, x_2, ...)$$

---

[1] vollstetig = completely continuous.

wird sobald man ε_1, ε_2, ... irgend solche Wertsysteme $\varepsilon_1^{(h)}$, $\varepsilon_2^{(h)}$, ... durchlaufen läszt, dasz einzeln

$$\underset{h=\infty}{L} \varepsilon_1^{(h)} = 0, \quad \underset{h=\infty}{L} \varepsilon_2^{(h)} = 0, \; ...$$

ist; dabei sind die Variablen stets an die Ungleichung $x_1^2 + x_2^2 + ... \leqq 1$ gebunden.

This is the point where the sequences (x_i) such that $\sum_i x_i^2 < \infty$ appear to be important, although there is no indication of a linear family (space).

Hilbert applies this notion to bilinear forms

$$B(x, y) = \sum_{i, k = 1}^{\infty} a_{ik} x_i y_k$$

and the corresponding 'Abschnitte'

$$\sum_{i, k = 1}^{n} a_{ik} x_i x_k.$$

B is called *vollstetig* if

$$B(x, y) = \lim_{n \to \infty} \sum_{i, k = 1}^{n} a_{ik} x_i y_k$$

uniform for all $x = (x_i)$, $y = (y_i)$ satisfying the conditions

$$\sum_{i=1}^{\infty} x_i^2 \leqq 1, \; \sum_{i=1}^{\infty} y_i^2 \leqq 1.$$

The kind of convergence of the variables in this definition is now called the *weak convergence*; note that Hilbert did not use this terminology.

Weak convergence in the Hilbert space l^2 is defined as follows:
The sequence $(x^{(n)})_{n \in N}$, $x^{(n)} \in l^2$, is said to converge weakly to $x \in l^2$ if

(i) $\|x^{(n)}\| \leqq M$ *for* $n = 1, 2, ...$,

(ii) $\lim_{n \to \infty} x_i^{(n)} = x_i$ *for* $i = 1, 2, ...$

With this definition of weak convergence it can be shown that the condition expressing that the bilinear form B is vollstetig is equivalent to the condition

$$\lim_{n \to \infty} B(x^{(n)}, y^{(n)}) = B(x, y)$$

for all sequences $(x^{(n)})$, $(y^{(n)})$ which are weakly convergent to x resp. y.

Weak convergence should well be distinguished from *strong convergence*, which is defined by means of the norm: *The sequence* $(x^{(n)})_{n \in N}$ *converges strongly to x if*

$$\lim_{n \to \infty} \sum_{k=1}^{\infty} |x_k - x_k^{(n)}|^2 = 0.$$

Any sequence which is strongly convergent is weakly convergent, but the converse is not true.

Example. The sequence defined by

$$e_1 = (1, 0, ...)$$
$$e_2 = (0, 1, ...)$$
$$\cdot \ \cdot \ \cdot \ \cdot \ \cdot \ \cdot \ \cdot \ \cdot$$
$$e_i = (0, 0, ..., 1, 0, ...)$$
$$\cdot \ \cdot \ \cdot \ \cdot \ \cdot \ \cdot \ \cdot \ \cdot$$

is weakly convergent to $(0, 0, ...)$ but it is not strongly convergent.

Remarks. 1. The second condition in the definition of weak convergence of a sequence $(x^{(n)})$ expresses the convergence of every sequence of coordinates $(x_i^{(n)})$. Note that the first condition, which expresses the boundedness of the sequence of the norms, is not a consequence of the convergence of the coordinates. One can even construct examples of sequences satisfying condition (ii) for which the sequence of the norms is unbounded. See [70], II, p. 25.

2. A function $f \colon l^2 \to \mathbb{R}$ which is continuous with respect to weak convergence is continuous with respect to strong convergence, but the converse is not true. That is to say: weak continuity is a stronger condition than strong continuity. This justifies the term 'vollstetig'.

3. The theorem of Fischer-Riesz shows the relation between the theory of functions of an infinite number of variables, such as Hilbert studied, and the theory of functionals ('fonctions de ligne'), which, as we have seen, was an important subject in the development of function-

al analysis. Indeed, any function of L^2 is uniquely determined by an element $(a_k) \in l^2$ and thus the functionals defined on L^2 correspond to the functions of the infinite number of variables a_k ($k = 1, 2, ...$).

Hilbert (*Grundzüge*, p. 127) also defined continuity (Stetigkeit). This definition runs:

Es heisze allgemein eine Funktion $F(x_1, x_2, ...)$ der unendlich vielen Variablen $x_1, x_2, ...$ an der Stelle $a_1, a_2, ...$ *stetig* wenn der Wert von $F(a_1 + \varepsilon_1, a_2 + \varepsilon_2, ...)$ gegen den Wert von $F(a_1, a_2, ...)$ konvergiert sobald die Summe der Quadrate der Gröszen $\varepsilon_1, \varepsilon_2, ...$ nach Null abnimmt, d.h. wenn

$$\underset{\varepsilon_1{}^2 + \varepsilon_2{}^2 + \cdots = 0}{L} \quad F(a_1 + \varepsilon_1, a_2 + \varepsilon_2, ...) = F(a_1, a_2, ...).$$

This is preceded by the restriction that only values of the variables are considered such that $(x, x) \leqq 1$. This notion is evidently based on strong convergence of the variables.

I already remarked that Schmidt ([113], p. 58) used the term 'starke Convergenz'. It seems that a definition of weak convergence was given for the first time by Riesz in his paper from 1910 in which he treated the spaces L^p ('die Klasse $[L^p]$'). In [107] p. 465 he wrote in the paragraph entitled 'Starke und schwache Konvergenz in bezug auf den Exponenten p':

Die Folge $\{f_i(x)\}$ von Funktionen der Klasse $[L^p]$ konvergiert in bezug auf den Exponenten p schwach gegen die Funktion $f(x)$ derselben Klasse, wenn
a) die Integralwerte

$$\int_a^b |f_i(x)|^p dx$$

insgesamt unterhalb einer endlichen Schranke liegen;
b) für alle Stellen $a \leqq x \leqq b$

$$\lim_{i = \infty} \int_a^x f_i(x)dx = \int_a^x f(x)dx$$

ausfällt.

In a footnote (p. 465) Riesz refers to Hilbert:

Für $p = 2$ steht der hier eingeführte Begriff der Schwachen Konvergenz in engem Zusammenhange mit jenem Konvergenzbegriffe, dessen sich Hilbert zur Definition der *vollstetigen* Funktionen von unendlich vielen Veränderlichen bedient; indem nämlich der schwachen Konvergenz einer Folge von Funktionen jene Hilbertsche Konvergenz ihrer Fourierschen Konstanten entspricht.

Riesz defined 'starke Konvergenz' by the relation

$$\lim_{i \to \infty} \int_a^b |f(x) - f_i(x)|^p \mathrm{d}x = 0$$

and for $p = 2$ he refers to the 'convergence en moyenne' as defined by E. Fischer ([107], p. 464).

Remarks. 1. In his fundamental book *Théorie des opérations linéaires* [4] Banach states (p. 135) with regard to weak convergence ('convergence faible') in L^p that Riesz *proves* the relations a) and b) which I quoted above. This is not correct. Banach defined the notion of weak convergence by means of the *dual space*; this was possible for him because the theory in his book was based on the axiomatic definition of normed spaces, which was known at this moment. In that case the conditions a) and b) appear as properties. But Riesz used a) and b) as a kind of *definition*. This was quite natural because in 1910–the year of Riesz's definition–the general notion of the dual space of a normed space had not yet been introduced. The dual space appeared after the theorem of Hahn-Banach, which proved the existence of non trivial continuous linear functionals; this was about 1927.

2. Note that Riesz did not introduce the concepts of weak and strong *topology*.

In his book from 1913 [109] p. 96, Riesz introduced similar notions for transformations; for instance a 'substitution A' is called

compIètement continue lorsqu'elle fait correspondre à chaque suite $(x_k^{(n)})$ $(n = 1, 2, \ldots)$, tendant de n'importe quelle façon vers un élément limite (x_k), une suite $(x_k^{(n)'})$ qui tend *fortement* vers l'élément (x_k').

In our modern language we speak of compact operators, i.e. operators that transform bounded sets into relative compact sets.

Riesz continued his research on this domain in an important paper from 1918 'Über lineare Funktionalgleichungen' [110]. Here Riesz developed the modern theory of linear operators by *direct methods*, i.e. without the limit procedure, as, for instance, Hilbert did; compare also Schmidt (1908, [113]). The theory is developed in the space $C(I)$ of real continuous functions defined on the interval I. But Riesz observes (p. 71):

> Die in der Arbeit gemachte Einschränkung auf stetige Funktionen ist nicht von Belang. Der in den neueren Untersuchungen über diverse Funktionalräume bewanderte Leser wird die allgemeinere Verwendbarkeit der Methode sofort erkennen; er wird auch bemerken, dasz gewisse unter diesen, so die Gesammtheit der quadratisch integrierbaren Funktionen und der Hilbert'sche Raum von unendlich vielen Dimensionen noch Vereinfachungen gestatten, während der hier behandelte scheinbar einfachere Fall als Prüfstein für die allgemeine Verwendbarkeit betrachtet werden darf.

Several of the notions, important for the axiomatic theory, are defined. 'Die zu Grunde gelegte Gesammtheit werden wir der Kürze halber als *Funktionalraum* bezeichnen.' The norm $\|f\|$ of the function f (Riesz writes $f(x)$) is defined as the maximum of $|f(x)|$ for $x \in I$, and he states (p. 72)

$$\|cf(x)\| = |c| \cdot \|f(x)\|; \quad \|f_1 + f_2\| \leqq \|f_1\| + \|f_2\|.$$

Then linear transformations (p. 72):

> Eine Transformation T, die jedem Elemente f unseres Funktionalraumes ein eindeutig bestimmtes Element $T[f]$ zuordnet, soll dann linear heiszen, wenn sie *distributiv* und *beschränkt* ist. Die Transformation heiszt distributiv, wenn identisch für alle f

$$T[cf] = cT[f], \quad T[f_1 + f_2] = T[f_1] + T[f_2]$$

> ist. Beschränkt heiszt die Transformation dann, wenn es eine Konstante M gibt derart, dasz für alle f

$$\|T[f]\| \leqq M\|f\|$$

> ausfällt.

The concept of a compact sequence is fundamental. Riesz gives the definition (p. 73): 'Eine Folge $\{f_n\}$ heisze nach Fréchet kompakt, wenn jede Teilfolge derselben eine gleichmässig konvergente weitere Teilfolge enthält.' This is used in defining 'vollstetige Transformationen'. I quote (p. 71):

Dieser Begriff gestattet eine besonders einfache und glückliche Formulierung der Definition der vollstetigen Transformation, die im wesentlichen einer ähnlichen Begriffsbildung von Herrn Hilbert für Funktionen von unendlich vielen Veränderlichen nachgebildet ist. (...)

Wir erklären nun: eine lineare Transformation heisze vollstetig, wenn sie jede *beschränkte* Folge in eine *kompakte* überführt. (p. 74).

As an example of 'vollstetige Transformationen' he mentions

$$T[f] = \int_a^x f(x)\mathrm{d}x,$$

$$K[f] = \int_a^b K(x, y)f(y)\mathrm{d}y$$

and as an example of a transformation that is not 'vollstetig' he mentions the identical transformation E (which transforms any bounded non-compact sequence into itself).

Finally he introduced as basic for his theory the concept of a 'lineare Mannigfaltigkeit' which he defines as follows (p. 74):

Darunter verstehen wir jede Mannigfaltigkeit von Elementen unseres Funktionalraumes, die folgenden Bedingungen genügt:
1) mit f, f_1, f_2 zugleich sind auch $cf, f_1 + f_2$ darin enthalten;
2) sind die Elemente einer gleichmäszig konvergenten Folge $\{f_n\}$ darin enthalten, so ist es auch die Grenzfunktion f.

With these tools Riesz treats the theory of the systems

$$T[f] - \lambda f = g,$$

and he obtains the results of the Fredholm theory of integral equations, formulated in terms of the dimension of the linear spaces of the solutions. One recognizes in this theory the modern theory of linear operators.

This paper, the source of which was the theory of linear integral equations, was fundamental for the development of the theory of linear operators.

I have to mention here an article of Helly, published in 1921, entitled 'Über Systeme linearer Gleichungen mit unendlich vielen Unbekannten' [65] because in this paper Helly was very close to the theorem of Hahn-Banach, a theorem implying the existence of non-trivial linear continuous functionals. The following paragraph shows the motivation for his work (p. 60):

> Die Bedingungen der Lösbarkeit eines Systems von unendlich vielen linearen Gleichungen mit unendlich vielen Unbekannten wurden, besonders durch die Arbeiten von E. Schmidt und F. Riesz, für den Fall aufgestellt, dasz die Koeffizienten und Lösungen gewissen Ungleichungen zu genügen haben. In der vorliegenden Arbeit soll gezeigt werden, dasz der wesentliche Inhalt der betreffenden Bedingungen darin liegt, dasz im Raum von abzählbar unendlich viel Dimensionen, in welchem die geometrische Interpretation des Gleichungssystems vor sich geht, eine Abstandbestimmung vorliegt, für welche das 'Dreiecksaxiom' gilt, oder um den Zusammenhang mit den Minkowskischen Begriffsbildungen hervorzuheben, der 'Aichkörper' ein konvexer Körper ist.

The geometry in the infinite dimensional space is introduced as follows (p. 66):

> Als Punkt x eines Raumes R_ω von abzählbar unendlich viel Dimensionen sei jede Zahlenfolge $x_1, x_2, x_3, \ldots$ bezeichnet, wobei die Gröszen x_k beliebige reelle oder komplexe Zahlen sein können. Der Punkt mit den Koordinaten $x_1, x_2, \ldots, x_n, 0, 0, 0, \ldots$ soll 'n^{ter} Abschnitt' des Punktes x genannt und mit $x|_n$ bezeichnet werden. Die Gesammtheit aller Punkte, für welche die Reihe
>
> $$(u, x) = \sum_{k=1}^{\infty} u_k x_k$$
>
> konvergiert und für welche die Gleichung
>
> $$(u, x) = c$$
>
> besteht, bilden eine lineare Mannigfaltigkeit, die mit $R_{\omega-1}$ bezeichnet werden soll. Entsprechend erfüllen die Punkte, die einem System van p linear unabhängigen Gleichungen genügen, einem $R_{\omega-p}$. Die Begriffe Gerade, Strecke, Ebene lassen sich ohne weiteres übertragen.

The concept of a distance (we now say a norm) is defined in an axiomatic way (p. 67):

Es sei nun eine Abstandsfunktion $D(x)$ gegeben, die jedem Punkte x eines gewissen Bereiches eine reelle positive Zahl zuordnet und die folgenden Bedingungen genügt:

I. Wenn x dem Definitionsbereich von $D(x)$ angehört, so soll auch λx ihn angehören und es soll

$$D(\lambda x) = |\lambda| D(x)$$

sein.

II. Wenn x und y dem Definitionsbereich von $D(x)$ angehören, so soll auch $x+y$ ihm angehören und es soll

$$D(x+y) \leqq D(x) + D(y)$$

sein.

III. Aus $D(x) = 0$ folgt $x = 0$.

Mit Hilfe der Funktion $D(x)$ kann jetzt der Begriff der Grenze definiert werden: Es sei $x = \lim_{v \to \infty} x^{(v)}$ wenn $D(x - x^{(v)})$ durch genügend grosse Wahl von v beliebig klein gemacht werden kann. Die Begriffe Umgebung, Häufungsstelle, Ableitung einer Punktmenge ergeben sich in bekannter Weise.[1]

There is a beginning of the concept of the dual space in this paper (p. 67):

Wie bei endlicher Dimensionalzahl, so läszt sich auch hier der Begriff der polaren Funktion definieren. Es sei $u = (u_1, u_2, u_3, \ldots)$ so beschaffen, dasz (u, x) für alle x mit endlichem $D(x)$ konvergiert, und dasz die obere Grenze der Werte von (u, x), wenn x auf das Gebiet $D(x) = 1$ beschränkt wird, die so bezeichnet werden soll

$$\overline{|(u, x)|}_{D(x) = 1},$$

endlich ist. Dann setzen wir

$$\Delta(u) = \overline{|(u, x)|}_{D(x) = 1}.$$

---

1 Compare the definition of Schmidt of $\|A\|$ in [113; 1908]. Helly's definition differs essentially from this definition of Schmidt, because Helly introduced the concept of a norm (without using the word) in an *axiomatic* way.

Und es ergibt sich, wie in § 1, für beliebige x die fundamentale Ungleichung

$$|(u, x)| \leqq \Delta(u)D(x).$$

Die so bestimmte Funktion $\Delta(u)$ genügt der Bedingungen I und II.

Now we define Δ by

$$\sup_{x \neq 0} \frac{|\langle u, x \rangle|}{\|x\|}.$$

The problem of the existence of a solution of a system of linear equations leads Helly to the following problem in which we recognize the problem of the existence of linear functionals (p. 74):

Es soll im Bereich der unendlich vielen Variablen $u_1, u_2, u_3, \ldots$ eine Operation $L(u)$ gesucht werden, die folgenden Forderungen genügt:

I. Die Forderungen der Linearität:

$$L(\lambda u) = \lambda L(u)$$
$$L(u' + u'') = L(u') + L(u'').$$

II. $|L(u)| \leqq M_1 \Delta(u)$

wobei M_1 eine beliebige Zahl $> M$ ist.

III. $L(a^{(v)}) = c_v \quad (v = 1, 2, \ldots).$

Kann eine Operation, die den gestellten Bedingungen genügt, gefunden werden, und kann dann ferner gezeigt werden, dass sich diese Operation in der Form

$$L(u) = (p, u), \quad D(p) \leqq M_1$$

darstellen läszt, so ist $x = p$ die gesuchte Lösung von (1). This system (1) is

$$(a^{(v)}, x) = c_v \quad (v = 1, 2, 3, \ldots),$$

satisfying the inequalities

$$\sum_{v=1}^{n} \lambda_v c_v \leqq M \Delta (\sum_{\gamma=1}^{n} \lambda_v a^{(v)})$$

for arbitrary n and λ_v, and where $a^{(v)}$ is a point of the sequence space.
Helly showed the existence of the linear operator L but he states that L cannot always be represented in the form $L(u) = (u, q)$. 'Wohl aber existiert immer eine Darstellung

$$L(u) = \lim_{v \to \infty} (u, q^{(v)}).'$$

(l.c. p. 80). Compare this with the representation of Hadamard (1903). Helly treats the special case in which the relation between D and $\varDelta$ is reflexive:

$$D(x) = \overline{|(u, x)|}_{\varDelta(u) = 1}.$$

This suggests duality.

All this shows that in this paper–which is almost forgotten–Helly was quite close to the theorem of Hahn-Banach, with this restriction that he only treated sequence spaces, and not the axiomatic theory.

Remark. In this paper Helly used the following interesting theorem on convex sets (l.c. p. 76):

Wenn in einem n-dimensionalen Raume ein System konvexer Körper gegeben ist, von welchen je $n+1$ mindestens einen Punkt gemein haben, so gibt es mindestens einen Punkt, der allen gegebenen Körpern angehört.

In a footnote Helly states that he already knew this property in 1913.[1]

It is most remarkable that an analogous property in infinite dimensional spaces–but then in the form of a condition on the space (the so called binary intersection property)–plays a decisive role in the theory of the extension of linear operators and in non-archimedean analysis (see [88], where a bibliography is to be found). I omit the details because they are not necessary for the purpose of this book.

The fundamental contributions of the preceding sections paved the way for a general theory of normed spaces and linear functionals and linear operators in an axiomatic form, but before entering the axiomatic era in the next section, a summary of the development which I sketched in the preceding pages will be useful.

[1] Radon proved this theorem in 1921 in a paper in the Math. Ann. Bd. 83, 113–115, 'Mengen konvexer Körper die einem gemeinsamen Punkt enthalten'. Helly communicated his theorem to Radon. See [26].

In the work of Hilbert, Riesz and Helly gradually rose the concept of a linear space and of linear functionals and operators. Especially Riesz and Helly gave definitions of these concepts, for instance under the name of 'lineare Funktionengebilde' or 'lineare Mannigfaltigkeit' thus giving the impression that these concepts were new or at least were not common tools of the mathematicians. Now the curious fact is that these notions were already introduced at the end of the 19th century even in its abstract axiomatic form by the Italian mathematicians I mentioned several times. But in the work of Hilbert, Riesz and Helly one hardly finds any reference to the work of Peano, Pincherle and Volterra. In Riesz's work from 1918 [110], p. 73, he refers to a paper by C. Arzelà, 'Sulle funzioni di linei' (*Memorie d. R. Acad. d. Scienze di Bologna*, serie 5, t. v (1895), p. 225–244). Hadamard referred to Volterra, Pincherle and Bourlet in his paper from 1903 [51]. As Riesz built on Hadamard's paper, one would be inclined to think that Riesz at least knew the work of the Italians. This raises the question: what could have been the reason that the concept of a linear space, axiomatically defined, became a more or less common tool in mathematics only in the years after 1920? The work of the Italian school seems to have been forgotten for many years, in spite of its importance. Did the theory of normed linear spaces, fundamental in modern mathematics, develop in the years after 1920 more or less independently from the results of the Italians or at least without much influence of their results?

In the chapters II and III I will discuss the work of the Italian school in more detail.

§ 5 *Axiomatics*

For the axiomatic theory of normed spaces the doctor's thesis of S. Banach, presented in June 1920 to the university of Léopol ($=$ Lwow) is fundamental; it was published in the journal *Fundamenta Mathematica* [2]. It is entitled 'Sur les opérations dans les ensembles abstraits et leur application aux équations intégrales'. Banach defines the notion of a normed linear space, in the form we use nowadays. He did not use the words 'linear space'; he wrote (l.c. p. 134):

Soit E une classe composée tout au moins de deux éléments, d'ailleurs arbitraires, que nous désignerons p.e. par $X, Y, Z, \ldots$.

The introduction to this paper gives an idea of Banach's aim (l.c. p. 133):

Introduction. L'opération c'est une relation univoque yRx c'est à dire, telle que

$$yRx \text{ et } zRx \text{ entraîne } y = z$$

pour tout x, y, z.

Chaque relation yRx comporte un *domaine* (c'est la réserve des y) et un *contre-domaine* (la réserve des x) ou *champ*. L'opération fonctionnelle ou la *fonction de ligne* c'est une opération dont le domaine et le contre-domaine sont des ensembles de fonctions. La notion de fonction de ligne fut introduite par M. Volterra. Des recherches à ce sujet ont été faits par M.M. Fréchet, Hadamard, F. Riesz, Pincherle, Steinhaus, Weyl, Lebesgue et par beaucoup d'autres. Dans les premiers ouvrages on admettait que le domaine et le contre-domaine sont des ensembles de fonctions continues admettant les dérivées d'ordres supérieurs. Ce ne furent que les travaux de Hilbert qui, bien qu'ils traitaient les formes quadratiques à une infinité de variables et non pas les fonctions de ligne, ont apporté des résultats susceptibles à être transferés facilement sur les théorèmes concernant les opérations dont le domaine et le contre-domaine se composent des fonctions de carré intégrable (L).

M. Wilkosz et moi, nous avons certains résultats (que nous nous proposons de publier plus tard) sur les opérations dont les domaines sont des ensembles de fonctions duhameliennes, c'est à dire qui sont les dérivées de leurs fonctions primitives.

L'ouvrage présent a pour but d'établir quelques théorèmes valables pour différents champs fonctionnels, que je spécifie dans la suite. Toutefois, afin de ne pas être obligé à les démontrer isolément pour chaque champ particulier, ce qui serait bien pénible, j'ai choisi une voie différente que voici: je considère d'une façon générale les ensembles d'éléments dont je postule certaines propriétés, j'en déduis des théorèmes et je démontre ensuite de chaque champ fonctionnel particulier que les postulats adoptés sont vrais pour lui.

Banach gives several examples: 'l'ensemble des fonctions continues, l'ensemble des fonctions sommables' etc. I omit them here. Then he gives the following axioms (l.c. p. 134):

Soit E une classe composée tout au moins de deux éléments, d'ailleurs arbitraires, que nous désignerons p.e. par $X, Y, Z, \ldots$.

a, b, c, désignant les nombres réels quelconques, nous définissons pour E deux opérations suivantes:

1) l'addition des éléments de E

$$X+Y, X+Z, \ldots$$

2) la multiplication des éléments de E par un nombre réel

$$aX, bY, \ldots$$

Admettons que les propriétés suivantes sont réalisées:

I_1] $X+Y$ est un élément bien déterminé de la classe E,
I_2] $X+Y = Y+X$,
I_3] $X+(Y+Z) = (X+Y)+Z$,
I_4] $X+Y = X+Z$ entraîne $Y = Z$,
I_5] Il existe un élément de la classe E déterminé θ et tel qu'on ait toujours $X+\theta = X$,
I_6] $a \cdot X$ est un élément bien déterminé de la classe E,
I_7] $a \cdot X = \theta$ équivaut à $X = \theta$ ou $a = 0$,
I_8] $a \neq 0$ et $a \cdot X = a \cdot Y$ entraînent $X = Y$,
I_9] $X \neq \theta$ et $a \cdot X = b \cdot X$ entraînent $a = b$,
I_{10}] $a \cdot (X+Y) = a \cdot X + a \cdot Y$,
I_{11}] $(a+b) \cdot X = a \cdot X + b \cdot X$,
I_{12}] $1 \cdot X = X$,
I_{13}] $a \cdot (b \cdot X) = (a \cdot b) \cdot X$.

Nous introduisons en même temps les définitions suivantes:

(a) $-X = (-1) \cdot X$

(b) $X-Y = X+(-1) \cdot Y$.

Toutes les règles algébriques des signes $+$, $-$ et de la multiplication (d'un élément de E par un nombre) restent valables pour un système E. Commes exemples d'un tel système peuvent servir: les vecteurs, les formes de Grassmann, les quaternions, les nombres complexes etc.

Admettons en suite que

II] Il existe une opération appelée norme (nous la désignerons par le symbole $\|X\|$), définie dans le champ E, ayant pour contre-domaine l'ensemble des nombres réels et satisfaisant aux conditions suivantes:

II$_1$] $\|X\| \geqq 0$,
II$_2$] $\|X\| = 0$ équivaut à $X = \theta$,
II$_3$] $\|a \cdot X\| = |a| \cdot \|X\|$,
II$_4$] $\|X + Y\| \leqq \|X\| + \|Y\|$,
III] Si 1° $\{X_n\}$ est une suite d'éléments de E
     2° $\lim\limits_{\substack{r=\infty \\ p=\infty}} \|X_r - X_p\| = 0$,

il existe un élément X tel que

$$\lim_{n=\infty} \|X - X_n\| = 0.$$

I observe that the concept of a norm was introduced about the same time by Helly [65] and Banach. In Banach [4] there are references to Helly.

As to III Banach says that $\{X_n\}$ converges in norm to X and he writes

$$\overline{\overline{\lim}}_{n=\infty} X_n = X.$$

Banach then studies the geometry in these spaces ('champs').

For instance for any given $X_1 \in E$ and any positive number r he defines the sphere $K(X_1, r)$ as the set of all $X \in E$ satisfying $\|X - X_1\| \leqq r$ and he obtains the result: 'Chaque sphère n'a qu'un seul centre et un seul rayon.'[1]

He defines additive operators and continuous operators. He proves the equivalence of boundedness and continuity of linear operators, a

---

i Note that this is not true in the so-called non-archimedean analysis, in which the axiom II$_4$] is replaced by the stronger inequality

$\|X + Y\| \leqq \max(\|X\|, \|Y\|)$.

In this case one easily verifies that any point of a sphere can serve as the centre of that sphere. See chapter IV § 5 for some consequences of this fact.

property that we already met in the work of Riesz and Helly, in the case of concrete spaces. But there is not yet a theorem on the existence of non-trivial linear continuous functionals on such a space. Banach and Hahn proved this theorem some years later.

Before proceeding with the treatment of this line of development I will make some remarks. The definition of a normed space which I quoted above is the same one as was given by Banach in his book *Théorie des opérations linéaires* [4]. In view of axiom III the spaces are complete normed linear spaces and in this book Banach calls them 'espaces-*B*', which later became the notion of a Banach space.

In the set of axioms I one recognizes the modern axioms of a vector space E over the field of the real numbers; Banach used the notation θ instead of 0, as we do nowadays, evidently to distinguish the zero element of the space from the number 0. Banach stated explicitly all these axioms and he still did so in his book *Théorie des opérations linéaires* which was published in 1932. It seems that the notion of a linear space, without a norm and as a notion in algebra, was not yet a common tool for the mathematicians in those years. Note, however, that this notion already appeared at the end of the 19th century with the Italian mathematicians I mentioned before several times. Banach refers in his work to the Italians (but not to Peano who gave a definition of a linear space, even infinite dimensional, in 1888). The situation asks for the study of the development of the notion of a linear space in algebra, a notion which must precede the notion of a normed space or a Banach space. We will perform this task in chapter II.

About the same time Wiener defined the notion of a normed linear space. In *Fundamenta Mathematica* t. IV he published a paper 'Note on a paper of M. Banach' [135], in which he repeats the axioms of Banach and gives some applications. On p. 143 of this article Wiener added the following footnote:

As a final comment on this paper, I wish to indicate that postulates not unlike those of M. Banach have been given by me on several occasions (Comptes Rendus of the Strasbourg mathematical conference of 1920; Proceedings of the London Mathematical Society, Ser. 2, Vol. 20, Part 5, pp. 332, 333, a forthcoming paper in the Bulletin de la Société Mathématique de France). However, as my work dates only back to August and September, 1920, and M. Banach's work was already presented for the degree of doctor of philosophy in June, 1920, he has the priority of original composition. I have employed M. Banach's postulates rather than my own because they are in a form more immediately adopted to the treatment of the problem in hand.

For this paper in the Bulletin de la Société Mathématique de France see [134]. Wiener is inspired by the work of Fréchet on general analysis, who started the study of the subject in 1904 (see [29], [32]). I quote from Wiener's paper:

The *calcul fonctionnel*, or the study of the limit properties of an abstract assemblage, has been investigated in the course of the last fifteen years from a number of distinct standpoints. In addition to the notion of sequential limit which furnished the startingpoint of Fréchet's well known thesis, and the more restrictive concept of *écart* (or *distance*, as Fréchet, now calls it), there is Riesz' nonsequential limit, which Fréchet, in turn, has discussed as a special case of an extremely general notion of neighborhood.[1]

If, however, we consider the *calcul fonctionnel* with reference to the obviously intimate bearing which it has on analysis situs–which has been defined essentially as the study of the invariants of the group of all bicontinuous biunivocal transformations–there is another avenue of approach to which our attention is immediately directed. This paper will be devoted to the discussion of the derivation of limit-properties from those of continuous transformations.

The investigation of this program led Wiener to the definition of a vector family–which is a concept analogous to the notion introduced by Banach.

We shall say that a set σ of entities is a *vector family* if there are associated with it operators $\oplus$, $\odot$, and $\| \ \|$ satisfying the following conditions:

(1) If ξ and η belong to σ, $\xi \oplus \eta$ belongs to σ,

(2) If ξ belongs to σ and n is a real number $\geqq 0$, $n \odot \xi$ belongs to σ,

(3) If ξ belongs to σ, $\|\xi\|$ is a real number $\geqq 0$,

(4) $n \odot (\xi \oplus \eta) = (n \odot \xi) \oplus (n \odot \eta)$,

---

1 See [31]. For general analysis and the work of Fréchet and Moore see chapter III.

(5) $(m \odot \xi) \oplus (n \odot \xi) = (m + n) \odot \xi,$

(6) $\|m \odot \xi\| = m \cdot \|\xi\|,$

(7) $\|\xi \oplus \eta\| \leq \|\xi\| + \|\eta\|,$

(8) $m \odot (n \odot \xi) = mn \odot \xi.$

Wiener gave several examples of vector families; I omit them because they are the common examples of such spaces. For more personal information about Wiener's work on normed spaces see his autobiography [136], p. 60.

The papers by Banach and Wiener were important steps in the development of modern analysis. The basic importance of their work is that they gave a synthesis in axiomatic form of the formal algebraic properties of a linear space and topological (metrical) properties sufficient to build an analysis. This makes it possible to apply algebraic methods to analysis. Notions such as linear subspaces, residue classes, quotient spaces and direct sums are frequently used in this analysis. One proves theorems for sets of functions, or sets of elements, rather than properties of the individual elements. The algebraization is even more apparent in the theory of normed algebras and, as a special case, function algebras. In a normed algebra there is, in addition to the structure of the space as an additive group, a not necessarily commutative multiplicative structure on the space, connected with the additive structure by the distributive laws, which makes the space into a ring; the norm is then supposed to satisfy the relation $\|xy\| \leq \|x\| \cdot \|y\|$. For some more information see chapter IV.

But functional analysis could only start its rapid development after the so called theorem of Hahn-Banach had been proved.

Fundamental work was done by H. Hahn in his paper 'Über lineare Gleichungssysteme in linearen Räumen', published in 1927 [54].

Hahn's starting point is in the theory of integral equations; his theory is formulated in normed linear spaces. I quote the introduction (l.c. p. 214):

Bekanntlich sind die Integralgleichungen zweiter Art:

$$\varphi(s) + \int_a^b K(s, t)\varphi(t)\mathrm{d}t = f(s)$$

der Untersuchung erheblich leichter zugänglich, als die Integral-gleichungen erster Art:

$$\int_a^b K(s, t)\varphi(t)\mathrm{d}t = f(s).$$

Es liegt das offenbar daran, daß wir die Auflösung der Gleichung $\varphi(s) = f(s)$ vollständig beherrschen, und durch Hinzutreten des Zusatzgliedes $\int_a^b K(s, t)\varphi(t)\mathrm{d}t$ die für die Gleichung $\varphi(s) = f(s)$ herrschenden durchsichtigen Verhältnisse nicht allzusehr gestört werden. Es liegt also die Frage nahe: sei in irgendeinem linearen Raume, dessen Punkte wir mit x bezeichnen, ein lineares Gleichungs-system gegeben:

$$u_y(x) = c_y,$$

von dem wir wissen, daß es auflösbar ist; unter welchen Umstän-den wird man daraus auf die Auflösbarkeit des linearen Gleichungs-systemes:

$$u_y(x) + v_y(x) = c_y$$

schließen können? Mit dieser Frage sollen sich die folgenden Zeilen beschäftigen. Als ausschlaggebend erweist sich dabei der in § 3 auseinandergesetzte Begriff der *Vollstetigkeit* des Systems der Li-nearformen $v_y(x)$ in bezug auf das System der Linearformen $u_y(x)$, der eine direkte Verallgemeinerung des von *Fr. Riesz* eingeführten Begriffes der Vollstetigkeit einer linearen Transformation darstellt. With regard to the linear system Hahn observes that in the case of integral equations x corresponds to the function φ and y takes all the values in the interval $[a, b]$. For the definition of a normed linear space Hahn refers to Banach [2] and to his paper 'Über Folgen linearer Operationen', published in 1922 [53] in which he gave the definition of this concept just like Banach did. There are also references to the work of Riesz and Helly. Hahn showed in this paper [54] the existence of continuous linear functionals $\neq 0$ on any complete normed space E over the field of the reals. He obtains this result by a procedure of extension of a linear functional f_0 defined on a linear subspace satisfying an inequality $|f_0(x)| \leqq M \cdot D(x)$, where D is the norm ('Massbestimmung') on the space. This is done by means of transfinite induction (a transfinite generalization of the ordinary induction process) with respect to the dimension (nowadays we use Zorn's lemma). In this paper Hahn already considered the set of all linear continuous functionals on the space E and he proved that this set is again a complete normed linear space.

Banach continued the work–referring to Riesz and Helly, but not to Hahn–in a paper from 1929 in which he proved a theorem on the extension of linear functionals in a more general situation. Instead of a norm the existence of a positive homogeneous function p, satisfying the triangle inequality, is supposed. I quote this theorem in the form Banach gave it in his book from 1932 ([4], p. 27). Let E be a linear space ('espace vectoriel') over $\mathbb{R}$.

Etant données

1° une fonctionnelle $p(x)$ définie dans E et telle que l'on ait pour tous x et y de E

$$p(x+y) \leqq p(x)+p(y) \text{ et } p(tx) = tp(x) \text{ pour } t \geqq 0,$$

2° une fonctionnelle additive et homogène $f(x)$ définie dans un espace vectoriel $G \subset E$ (bien entendu, avec les mêmes définitions des opérations fondamentales) et telle que l'on ait pour tout $x \in G$

$$f(x) \leqq p(x),$$

il existe toujours une fonctionnelle additive et homogène $F(x)$ définie dans E et telle que l'on a

$$F(x) \leqq p(x) \text{ pour tout } x \in E \text{ et } F(x) = f(x) \text{ pour tout } x \in G.$$

Observe that in this theorem no topology is presupposed to be defined on the space E–as is the case in normed spaces (Banach did not use concepts of topology in his book from 1932). In this form it is rather an algebraic theorem. Afterwards it was called the *theorem of Hahn-Banach*. This theorem implies again the existence of continuous linear functionals $\neq 0$ in the case of normed linear spaces [4].

More precisely: if E is a normed linear space one has the following well known theorem:

For any $x_0 \in E$ there exists a linear continuous functional F on E such that

$$F(x_0) = \|x_0\|, \quad \|F\| = 1.$$

With Banach's book *Théorie des opérations linéaires* functional analysis began its course. Starting with the theory of additive groups and of metric (not necessarily normed) spaces, Banach develops the theory of normed linear spaces ('espaces-B'). It contains a wealth of ideas; it is a treatment of linear analysis. Many of the concepts of modern functional analysis are introduced in this book. One finds properties of the dual space–that is the space of continuous linear functionals–, relations between a normed space and its dual space, the form of the linear functionals in special spaces, where one meets again the representation theorem of Riesz. There are also the *uniform boundedness theorem of Banach-Steinhaus*, the *open mapping theorem of Banach*, which are, together with the theorem of Hahn-Banach, three fundamental theorems for application and further development of functional analysis (see chapter IV, § 7). There is a theory of linear operators with applications on integral equations, developed by Riesz [110] in the space $C(I)$, but now presented for B-spaces, further a theory on infinite systems of linear equations. But there are also unexpected applications, such as on the existence of continuous nowhere differentiable functions and on the existence of continuous functions whose Fourier series is divergent in a point. The book concludes with a list of unsolved problems. Banach's book marked an epoch in the development of modern analysis. It was followed by many papers, published especially in the journals *Fundamenta Mathematica* and *Studia Mathematica*.

The axiomatic theory of the Hilbert space was provided in 1930 by J. von Neumann [94].[1] The spaces l^2 and L^2 are models, that is special realizations, of the Hilbert space; frequently other models are used in analysis. In this paper von Neumann characterized Hilbert space by means of 5 postulates called A, B, C, D, E (linear space, existence of

---

[1] Some earlier work of von Neumann dates from 1927. For information on the development see [120], p. III–VI.

an inner product, separability, existence of arbitrary many linearly independent elements, completeness). Concerning the axioms A and B von Neumann refers to H. Weyl:

Die Bedingungen A, B sind übrigens in Anlehnung an ein (für endlich viele Dimensionen aufgestelltes) Axiomensystem des Vektorraumes von H. Weyl (*Raum, Zeit, Materie*, 5e Auflage, Berlin 1923, §§ 2–4) gebildet.

As the introduction of the concept of 'linear space' deserves special attention, we shall devote the next chapter to it.

Later on the separability (that is the existence of a dense countable set of elements) was dropped. For completeness sake we now give the modern definition of a Hilbert space:

A Hilbert space over the field of the reals is a linear space H over the reals on which is defined an inner product and on which a norm $\|\cdot\|$ is defined by

$$\|x\| = (x, x)^{\frac{1}{2}}.$$

H is complete with respect to the metric defined by the norm. Hilbert space over the complex numbers is defined in an analogous way; the inner product satisfies the relation $(x, y) = \overline{(y, x)}$, where by $\bar{z}$ is denoted the complex conjugate of the complex number z.

In the preceding paragraph I gave a survey of the great lines of the development until Banach's book. I can only give a rough sketch of the further development. Many results were obtained in the theory of Banach spaces. In particular the Polish school, with names as Steinhaus, Saks, Mazur, Borsuk, has contributed very much to this domain of mathematics. For information on the Polish school see [140].

One of the lines of the development led to the study of *non-normed spaces*. This way followed under the influence of the increasing importance of topology. The concept of a general topological space was already known for many years. Hausdorff, for instance, formulated the axioms of a topological space in 1914 in his book *Grundzüge der Mengenlehre* [58]. Already in 1906 Fréchet defined the notion of a distance on an abstract set. In 1926 he studied linear spaces on which a certain topology is defined and he called these spaces topologically affine. I refer the reader to Fréchet [32], in which book there is an extensive bibliography. In 1934 Kolmogoroff defined the concept of a linear topological space [76]. *A linear topological space* is a linear space which is at the same time a topological space such that the algebraic structure of the space and the topology are connected with each other by the condition that addition and scalar multiplication are continuous operations. That is the operations $(x, y) \to x + y$ and $(x, a) \to ax$ are continuous. Normed spaces are linear topological spaces, but the converse is not true. Kolmogoroff gave a necessary and sufficient condition for a linear topological space to be isomorphic to a normed space. This condition is formulated in terms of convex sets, already studied by Helly. In this paper Kolmogoroff defined the concept of a bounded set (in the non-metric case).

However the structure of a linear topological space is not strong enough to have a theory based on it. In 1935 von Neumann defined the so called *locally convex spaces*, which are linear topological spaces in which the topology is determined by a basis of neighborhoods of 0 which are convex sets.[1] Such a space is not necessarily a normed space but, as a consequence of the fact that there is a basis of convex neighborhoods, the topology can be described by a system of seminorms.[2]

The locally convex spaces have shown to be very useful in analysis. The theorem of Hahn-Banach is valid in these spaces and so the dual space can be defined. Eventually a vast theory of these spaces was developed in which the theory of duality–that is the theory of the relations between the space and its dual–takes an important place. Various kinds of topologies (strong topology, weak topology) appear now under a new light (see Dieudonné [23]).

For a bibliography of the older literature on linear topological spaces see [69].

One interesting example of applications of locally convex spaces is the *theory of distributions*, due to the French mathematician L.

---

[1] Von Neumann defined also linear topological spaces; there is no reference to Kolmogoroff.

[2] A semi norm on the space E is defined as mapping p from E into $\mathbb{R}$ such that $p(x) \geqq 0$ for all $x \in E$, $p(0) = 0$, $p(ax) = |a| p(x)$, $p(x+y) \leqq p(x) + p(y)$. Semi norms (although not under this name) were already introduced by Minkowski in his *Geometrie der Zahlen*.

Schwartz. The so called distributions–some authors call them generalized functions–are linear functionals defined on certain locally convex spaces whose elements are functions which have derivatives of any
order. The concept of a derivative of a distribution is introduced in an
appropriate way: it is shown that a distribution has derivatives of any
order. The theory of distributions is very important in the modern
theory of partial differential equations (see [138]).

A second line in the development of functional analysis led to the
theory of *normed algebras*. Normed algebras were introduced by
Gelfand in 1941 [40]. They turned out to be of considerable importance
and there is a vast literature on this subject. I shall give some examples
in chapter IV.

§ 6 *Operator theory*

In the preceding pages I wrote several times on functional operations
as mappings of one function space into another and I treated in some
detail the theory of linear functionals, that are mappings into ℝ. In
the beginning the operations appeared as transformations or substitutions. In a later phase linear operators were defined in a general way
in Banach spaces.

There is still another way by which operators came into mathematics; I will give a short sketch.

This introduction is much older than the development that I described in the preceding sections and it has a very different character.
The theory of operators which I mean here is a formal method for
calculating with certain symbols; it is a calculus for operators, at least
in the beginning without geometrical interpretations. This will become
clear from an example.

The use of symbolical methods and of a formal apparatus goes far
back. Already Leibniz occupied himself with formal algorithms.
Lagrange studied the algebraization of infinitesimal calculus. He
introduced an operator $\varDelta$ and formed an algebraic calculus with this
operator:

$$(\varDelta f)(x) = f(x+\alpha) - f(x),$$

introducing, for instance Δ^2, Δ^3, ... but also Δ^{-1}.

I mention Lagrange's *Théorie des fonctions analytiques*, published in 1797. The complete title shows the tendency towards algebraization: *Théorie des fonctions analytiques, contenant les principes du calcul différentiel, dégagés de toute considération d'infiniments petits ou d'évanouissans, de limites ou de fluxions, et réduits à l'analyse algébrique des quantités finies.* Here differentiation is defined in a formal way by means of power series.

Formal methods were especially developed in England in the 19th century. Names of mathematicians who worked on this domain are Gregory, De Morgan, Boole and in particular I mention Heaviside.

Heaviside was led to a formal calculus with operators from the side of electrotechnics (about 1887). I illustrate his method by an example which I borrowed from [38]; for historical details see that paper.

Example. Consider the following differential equation, which is met in the theory of electricity:

$$L\frac{dI}{dt} + RI + \frac{1}{c}\int Idt = E.$$

Suppose

$E = 0$ for $t < 0$,

$E = 1$ for $t \geqq 0$.

Heaviside considers d/dt as an operator D. The operator D^{-1} is then evidently an integral operator. With these notations the equation becomes

$$\left(LD + R + \frac{1}{c}D^{-1}\right)I = E.$$

Solving I one finds

$$I = \frac{E}{LD + R + \frac{1}{c}D^{-1}} = \frac{DE}{LD^2 + RD + \frac{1}{c}}.$$

Now, the member at right must be given a meaning. For adequate values of α, β, γ one finds

$$I = \gamma \left(\frac{D}{D - \alpha} - \frac{D}{D - \beta} \right) E,$$

and, writing the fractions as a power series in D^{-1},

$$I = \gamma \left(\sum_0^\infty \alpha^n D^{-n} - \sum_0^\infty \beta^n D^{-n} \right) E.$$

Evidently D^{-n} must be given the meaning of an n-fold integral of E, that is $t^n/n!$. So one finds the solution

$$I = \gamma \left(\sum_0^\infty \frac{(\alpha t)^n}{n!} - \sum_0^\infty \frac{(\beta t)^n}{n!} \right) = \gamma (e^{\alpha t} - e^{\beta t}).$$

One may verify that this is indeed a solution.

Heaviside has given many examples of this method, but nowhere he gives a justification of his formal apparatus. He was never troubled by divergent series.

This is a kind of algebraization of analysis quite different from what I described in the preceding sections. It is a symbolic method without giving insight in what is going on.

About 1925 it was observed that Heaviside's apparatus is connected with the so called Laplace-transformation. Let f be a function defined for $t \geq 0$. The Laplace-transformation associates to f a function Lf (provided that the integral exists) defined by

$$(Lf)(s) = \int_0^\infty e^{-st} f(t) \mathrm{d}t.$$

Applied to the derivative f' one finds (under adequate conditions)

$$(Lf')(s) = \int_0^\infty e^{-st} f'(t) \mathrm{d}t = s \int_0^\infty e^{-st} f(t) \mathrm{d}t,$$

that is

$$(Lf')(s) = s(Lf)(s).$$

Thus the operation of differentiation is reduced to multiplication.

Heaviside's method can be justified by means of this transformation but this requires the apparatus of classical analysis; it is beyond the scope of this book to give the details (see [25]). In the form of the

Laplace-transformation we are nearer to the subjects of the preceding sections, but the profit of short symbolical methods has then gone lost. For more information on the development of operator theory in this period I refer the reader to the Encyclopedia-paper of Pincherle (1905; [100]).

In more recent years (about 1950) Mikusinski returned to operator theory. He considers the commutative ring of the continuous complex valued functions on $(0, \infty)$. The element of the ring, associated to the function f is designed by $\{f(t)\}$. Addition is defined as usual by

$$\{f(t)\} + \{g(t)\} = \{f(t) + g(t)\}.$$

Multiplication is defined by convolution:

$$\{f(t)\} \cdot \{g(t)\} = \{\int_0^t f(t-\tau)g(\tau)\mathrm{d}\tau\}.$$

I take the following short survey of Mikusinski's method from Freudenthal's paper [25].

It follows from a theorem of Titchmarsh that this ring has no divisors of zero and so it can be embedded in a field (the quotient field) which we denote by C. It is in this field that operator theory is developed because now any element $\neq 0$ has an inverse (but one must be cautious; the inverse is not always a function). The function

$$l = \{1\}$$

corresponds to the integral operator because

$$\{1\}\{f(t)\} = \{\int_0^t f(\tau)\mathrm{d}\tau\}.$$

Writing

$$g = lf = \{\int_0^t f(\tau)\mathrm{d}\tau\},$$

one has by definition

$$f = l^{-1}g.$$

Instead of l^{-1} we will write D and it is easily proved that

$$l^{-1}\{f(t)\} = D\{f(t)\} = \{f'(t)\} + \{f(0)\},$$

and thus D is nearly the differential operator. This leads to an apparatus for solving differential equations more or less in the line of Heaviside, but it is somewhat isolated from the theory of operators in a Banach space which is in fact a theory of mappings. For Mikusinski's theory see [86].

The theory of linear operators in Banach spaces is the theory of linear transformations of a Banach space E into another F; also transformations of E into itself. The transformation is then defined by the operator. If T is the operator, one studies, for instance, transformations

$$Tx = y,$$

where $x \in E$, $y \in F$, or, if T is a transformation of E into itself, $x, y \in E$.

In his *Théorie des opérations linéaires* Banach already occupied himself with such problems (Chapitre x, Equations fonctionnelles linéaires) and he applied them to integral equations. The solution of an equation $Tx = y$ is evidently connected with the existence of an inverse T^{-1} of T.

There is now an extensive literature on operators in Banach spaces, especially in Hilbert spaces. I can only very roughly indicate some subjects. There is, for instance, the problem of eigenvalues of T, connected with the equation

$$(T - \lambda I)x = y,$$

where I is the identity transformation. This leads to the spectral theory of operators: the spectrum of T is the set of values of λ for which $T - \lambda I$ has no inverse. It is connected with the solution of the homogeneous equation $(T - \lambda I)x = 0$ (compare the Fredholm theory of integral equations). Banach already studied the spectrum. Several kinds of operators are studied: bounded (continuous) operators, but also unbounded operators; completely continuous (compact) operators; self-adjoint operators, connected with the dual space; and even non-linear operators. I refer the reader to [138], [139].

CHAPTER II

The development of the notion of a linear space

In this chapter I will deal with the history of the development of the notion of a linear space, a concept that is fundamental for the theory of normed spaces, locally convex spaces and for functional analysis (and of course for many other branches of mathematics, not belonging to the subject of this book). The theory of linear spaces belongs to *algebra*; there are no questions of convergence because no topology is defined on the space. The modern definition of a linear space reads:

Let K be a field. A linear space E over K is a set of elements satisfying the following conditions:

 (i) *E is an Abelian group under an operation denoted by* $+$;

 (ii) *There is a mapping from $K \times E$ into E such that for any $a, b \in K$, $x, y \in E$*

$$a(x+y) = ax+by,$$

$$(a+b)x = ax+by,$$

$$a(bx) = (ab)x,$$

$$1x = x.$$

I consider the case in which K is the field of real numbers, following the line of historical development. The crucial point is that the elements of the set E are not specified, in contrast to the more concrete situations handled by various authors. We have seen that Riesz and Helly considered sets of functions–continuous or integrable in some way–such that the properties (i) and (ii) are satisfied, and they called them 'totalité de fonctions', 'Klasse von Funktionen', 'lineare Mannigfaltigkeit'.

How did the abstract definition come into mathematics? In view of the fact that it took mathematics roughly a century to develop from the first ideas to the present axiomatic theory, a more or less

87

chronological treatment seems justified. First I observe that the notion of an n-dimensional geometry goes back to the first half of the 19th century. In a paper from 1843 Cayley studied n-dimensional analytical geometry ('Chapters in the analytical geometry of (n) dimensions'; Collected Math. Papers, Vol. 1, 55–62). I mention Riemann's 'Habilitationsvorlesung', 'Über die Hypothesen, welche der Geometrie zu Grunde liegen' (1854) in which the notion of an n-dimensional manifold appears. This paper was one of the sources of topology (Analysis Situs); several mathematicians (Betti, Poincaré) worked on the creation of this chapter of mathematics in the second half of the 19th century (see M. Bollinger, *Geschichtliche Entwicklung des Homologiebegriffs*; Inaugural-Dissertation Zürich 1972). However, this is not in my domain. See also [144].

The development of the theory of *linear spaces* is connected with several domains of mathematics. On the one hand the subject is connected with the development of algebra. For this aspect one should have to consult the work of all the mathematicians of this period who contributed to algebra: for instance Dedekind, Kronecker, Sylvester, Frobenius and Cayley. On the other hand there is the influence of the development of vector analysis in the traditional sense; in this field a vector is a geometric object, considered to have a length and a direction[1] (Hamilton [44]), and vectorial systems are studied. Here one should turn to Hamilton and his theory of quaternions (from 1843 onward), to the work of Gibbs; one can easily provide a long list of mathematicians in this field. The history of vector analysis in this sense was recently studied by Crowe [19]. As he remarks (p. 242), vector analysis in this traditional sense is only indirectly associated with the theory of linear spaces (vector spaces). Therefore I don't consider vector analysis in my survey. In particular I don't treat the work of Hamilton, although it was very important for the development of formal systems. For detailed information on this point I refer the reader to Crowe [19]. See also Hankel [56].

A survey of the relations with algebra is a domain of research for itself and is such an ambitious program that it cannot be dealt with

---

[1] Hamilton considered both vectors with a fixed initial point and free vectors for which he defined an equality relation (in present day mathematics we define equivalence classes).

in this book. I mention for instance the theory of matrices, which is closely connected with the theory of linear spaces.

In this chapter I confine myself to the more direct development of the theory of linear spaces; completeness is not pursued here. A more detailed study of the history of linear spaces, placed in the broad frame of algebra, would be highly interesting.

§ 1 *Bolzano*

There are some reasons to consider Bolzano as one of the first mathematicians who studied a geometry with undefined elements and the concept of a vector. In 1804 he published a book *Betrachtungen über einige Gegenstände der Elementargeometrie*, edited in Prague. It is reprinted in *Oeuvres de Bernard Bolzano*, edited in 1948 in Prague, in tome 5 *Mémoires géométriques*, p. 8–181. He is considering in it the foundations of elementary geometry and his aim is to give a rigorous treatment of the theory of triangles and parallel lines. The first chapter is entitled 'Versuch die ersten Lehrsätze von Dreiecken und Parallellinien mit Voraussetzung der Lehre von der geraden Linie zu erweisen', and the second one: 'Gedanken in Betreff einer künftig aufzustellenden Theorie der geraden Linie' *Oeuvres* t. 5, p. 37 [9]. In this chapter there are some germs of an axiomatic theory. Points, lines, planes are undefined elements for which Bolzano develops a theory. I quote some passages, which are rather philosophical:

Bevor der Geometer den Begriff der Gleichheit auf räumliche Dinge anwende, soll er erst die Möglichkeit gleicher räumlicher Dinge darthun. Vielleicht, dass sich hierzu der, I. Abth. § 19, aufgestellte Grundsatz[1] brauchen liesse, der in seiner Allgemeinheit also lautet: Wir haben von keinem bestimmten räumlichen Dinge (auch nur einem Puncte) eine Vorstellung a priori. (...)

Soviel ist, wie ich hoffe, ohne Widerspruch: daß der Begriff des Puncts – als eines bloßen Merkmals eines Raumes (σημεῖον), das selbst kein Theil des Raumes ist – in der Geometrie nicht entbehret

---

[1] This 'theorem' is (l.c. p. 18): 'Es ist uns keine besondere Vorstellung von irgend einer bestimmten Entfernung (oder absolute Länge der Linie), d.h. von einer bestimmten Art des Auseinanderseins zweier Puncte a priori gegeben.'

werden könne. – Dieser Punct ist allerdings ein bloß imaginärer Gegenstand, wie ich Hrn. Langsdorf gerne zugestehe. Auch Linie und Fläche sind dieß, und zwar alle drey noch in einem andern Sinne, als der geometrische Körper. Diesem nämlich kann in der Anschauung etwas adäquates gegeben werden, (und zwar alles, was in der Anschauung gegeben wird, ist Körper) nicht aber jenen. Und eben deshalb dürfte vielleicht jede versuchte reine Anschauung von Linien und Flächen (etwa durch die Bewegung eines Punctes) unmöglich seyn. Die in dieser Abhandlung versuchten Definitionen von der geraden Linie § 26, und der Ebene § 43 sind eben nach der Voraussetzung gemacht, daß beyde bloße Gedankendinge sind. (l.c. p. 38)

In the next chapter (l.c. p. 39) Bolzano considers couples of points and this may be seen as a slight anticipation on the notion of a vector as a system of two points. I quote this passage (l.c. p. 39):

Da Ein Punct für sich allein betrachtet nichts Unterscheidbares darbietet, indem wir von keinem eine bestimmte apriorische Vorstellung heben: so ist der einfachste Gegenstand der geometrischen Betrachtung ein System zweyer Puncte. Aus einem solchen Zugleichdenken zweyer Puncte entspringen gewisse Prädicate für dieselben (Begriffe), die bey der Betrachtung eines einzelnen Punctes nicht vorhanden waren. Alles was sich in der Beziehung dieser zwey Puncte aufeinander, und zwar in der Beziehung (...) des b auf a bemerken läßt, zerlege ich in zwey Theilbegriffe: I. Dasjenige, was dem Puncte b in Beziehung auf a so zukömmt, daß es unabhängig ist von dem bestimmten Puncte a (qua praecise hoc est et non aliud); was folglich auch in der Beziehung auf einen andern Punct, z.B. α gleich vorhanden seyn kann: genannt die Entfernung des Punctes b von a. II. Dasjenige, was dem Puncte b in Beziehung auf a so zukömmt, daß es abhängig ist bloß von dem bestimmten Puncte a; wovon nun getrennt werde, was schon in dem Begriffe der Entfernung liegt, d.h. was dem Puncte b auch in Rücksicht auf noch einen andern Punct zukommen kann: genannt die Richtung, in welcher b zu a liegt.

§ 7

Nun ist beyder Begriffe Möglichkeit zu zeigen. I. Der Entfernung. – Der bloße Begriff des Verschiedenseyns des Punctes *b* von *a* (des Auseinanderseyns) ist kein Theil des Begriffs der Beziehung des Punctes *b* auf *a* (des totius dividendi § 6); sondern wird dabey nothwendig schon vorausgesetzt. Soll *b* auf *a* bezogen werden; so muß die Vorstellung, daß *b* von *a* verschieden ist, schon vorausgehn. Um also die Realität des Begriffs der Entfernung, als eines Theilbegriffs von dem erwähnten Ganzen darzuthun, muß man beweisen, daß er mehr enthalte, als etwa das bloße Verschiedenseyn des *b* von *a*. Dieß thue ich so: Sollte der Begriff der Entfernung nichts Andres enthalten, so müßte der andere Begriff der Richtung das totum divisum noch ganz begreifen, d.h. der ganze Begriff der Beziehung *b* auf *a* müßte nichts enthalten, als was lediglich von dem bestimmten Puncte *a* abhängt; mit andern Worten: das System *ab* könnte durch das völlig bestimmt werden, was dem Punct *b* bloß abhängig von *a* zukömmt, also keinem andern Systeme zukommen kann; also hätten wir eine besondre apriorische Vorstellung von *a*, die wir von keinem andern Puncte hätten; welches gegen unsern Grundsatz ist. II. Der Begriff der Richtung kann nicht ganz leer seyn, weil sonst wieder der Begriff der Entfernung den eingetheilten Begriff ganz erschöpfen müßte. Aber ex definitione enthält dieser nur das, was dem Puncte *b* unabhängig von dem besondern Puncte *a* zukömmt, so daß er auch zukommen kann dem Systeme *b*α. Das einzutheilende Ganze aber enthält so viel, als dem Systeme *ab* nur allein zukömmt.

Bolzano developed a theory about 'distance' and 'direction', he introduced, for instance, the concept of opposite directions:

Man nenne (der Kürze wegen) der Punkt *m* (...) innerhalb oder zwischen *a* und *b*, wenn die Richtungen *ma*, *mb* entgegengesetzt sind. (...)

Ein Ding, welches alle jene, und nur jene Punkte enthält, die zwischen den zwei Punkten *a* und *b* liegen, heiszt eine gerade Linie zwischen *a* und *b*. (l.c. p. 44).

On the whole, this work may be seen as a forerunner to the work of Grassmann, although it anticipates perhaps more an axiomatizing of geometry. Operations with undefined elements are essential features in the development of linear spaces and in this respect it seems worthwhile to mention Bolzano here.

§ 2 *Laguerre*

In the further development towards the concepts of a linear space and of linear algebra I shall have to deal with the work of Laguerre, Grassmann, Peano and–then we have already advanced to the end of the century–Pincherle. The part of the work of these mathematicians that we are concerned with here–Laguerre, Peano and Pincherle published also on other domains of mathematics–must be seen against the background of some dissatisfaction that rose in the midst of the 19th century with the Cartesian method which was till then commonly used by the mathematicians. The Cartesian method was based on a set of arbitrary axes, in general without much connection with the problem itself. Geometry was thus reduced to algebra. This dissatisfaction led to attempts for finding *direct methods*, that is symbolic *coordinate-free* methods. In these attempts one finds the germs of linear algebra (see Kneebone [73], p. 141 a.f.).

I quote what Poincaré says about it in his introduction to the *Oeuvres de Laguerre*, I (1898), II (1905) ([79], p. v).

Dans le programme d'admission à cette Ecole (Ecole polytechnique), la place d'honneur appartient à la Géométrie analytique. Cette science se renouvelait par une révolution en quelque sorte inverse de la réforme cartésienne. Avant Descartes, le hasard seul, ou le génie, permettait de résoudre une question géométrique; après Descartes, on a pour arriver au résultat des règles infaillibles; pour être géomètre, il suffit d'être patient. Mais une méthode purement mécanique, qui ne demande à l'esprit aucun effort, ne peut être réellement féconde. Une nouvelle réforme était donc nécessaire: Poncelet et Chasles en furent les initiateurs. Grâce à aux, ce n'est plus ni à un hasard heureux, ni à une longue patience que nous devons demander la solution d'un problème, mais à une connaissance approfondie des faits mathématiques et de leurs rapports intimes. Les longs calculs d'autrefois sont devenus inutiles, car on peut le plus souvent en prévoir le résultat.

What Poincaré means here are the direct methods in geometry and the way of synthetic geometry. Laguerre (1834–1886) also contributed to this development; his first paper is from 1853.

Similar remarks were made by S. Lie in his 'Vorwort' to his book *Geometrie der Berührungstransformationen* [81]. There he sketched the development of analysis and geometry and their mutual influence in the course of time. I quote the following passage ([81], p. VI):

Unter allen Umständen ist es bedauerlich, dass die so grosse Entwickelung der Analysis in Deutschland in den letzten Jahrzehnten nicht von entsprechenden Fortschritten der Geometrie begleitet wurde. Hervorragende deutsche Geometer wie Möbius, Plücker, v. Staudt und Grassmann fanden zu ihrer Zeit nicht die richtige Würdigung von Seiten der dazu berufenen Stelle.

Die Zersplitterung der Mathematik hat, wie oben angedeutet wurde, auf die Vertreter der einzelnen Disciplinen oft eine ungünstige Wirkung geübt. Während nämlich einige Geometer so weit gehen, es als geradezu verdienstvoll zu betrachten, bei der Behandlung geometrischer Probleme auf die Hülfsmittel der Analysis vollständig, richtiger gesagt in möglichst grosser Ausdehnung zu verzichten, findet man wohl andererseits unter den Analytikern hier und da die Auffassung, dass die Analysis nicht allein unabhängig von der Geometrie entwickelt werden *könne*, sondern auch *müsse*, da nach ihrer Ansicht Beweise analytischer Sätze durch geometrische Betrachtungen nicht unbedingt zuverlässig sind.

In meinen wissenschaftlichen Bestrebungen bin ich immer von der Auffassung ausgegangen, dass es im Gegenteil wünschenswert ist, dass sich Analysis und Geometrie ebenso wie früher auch in unserer Zeit gegenseitig stützen und mit neuen Ideen bereichern. Diese Auffassung war im Jahre 1886 das Thema meiner Antrittsvorlesung an der Universität Leipzig.

The stream towards coordinate-free methods has its analogue in functional analysis, where it is met in the transition from special spaces – such as, for instance, the space l^2 defined as a sequence space – to the abstract definitions of a linear space. It can also be found in the transition from substitutions, defined as a matrix, towards linear operators. In the preceding pages one finds many examples which illustrate this situation.

As to Laguerre's place in the development of linear spaces, I mention his paper 'Sur le calcul des systèmes linéaires, Extrait d'une lettre adressée à M. Hermite' (1867 [79]).

In the introduction to the *Oeuvres de Laguerre* from which I quoted before, Poincaré observes that it is 'un Mémoire trop peu connu et dont la portée philosophique est très grande' (p. IX). The essence of the paper is that Laguerre developed a kind of linear algebra in it with applications on several parts of analysis. To give an idea of his method I quote some passages.

J'appelle, suivant l'usage habituel, *système linéaire* le tableau des coefficients d'un système de n équations linéaires à n inconnues. Un tel système sera dit *système linéaire d'ordre n* et, sauf une exception dont je parlerai plus loin, je le représenterai toujours par une seule lettre majuscule, réservant les lettres minuscules pour désigner spécialement les éléments du système linéaire.

Ainsi, par exemple, le système linéaire

$$\alpha \quad \beta$$
$$\gamma \quad \delta$$

sera représenté par la seule lettre majuscule A. Dans tout ce qui suit, je considérerai ces lettres majuscules représentant les systèmes linéaires comme de véritables quantités, soumises à toutes les opérations algébriques. Le sens des diverses opérations sera fixé ainsi qu'il suit.

Addition et soustraction. Soient deux systèmes de même ordre A et B; concevons que l'on forme un troisième système en faisant la somme algébrique des éléments correspondants dans chacun des deux premiers systèmes. Le système résultant sera dit la somme des systèmes A et B, et si on le désigne par C, on exprimera le mode de relation qui le rattache aux systèmes A et B par l'équation $C = A + B$. Si, par exemple, on a

$$A = \begin{matrix} a & b \\ c & d \end{matrix}, \qquad B = \begin{matrix} \alpha & \beta \\ \gamma & \delta \end{matrix},$$

il viendra

$$C = \begin{matrix} a+\alpha & b+\beta. \\ c+\gamma & d+\delta \end{matrix}$$

La soustraction sera évidemment définie d'une manière semblable. (l.c. p. 221).

Multiplication is defined as usual (l.c. p. 222): if A and B are defined as above, the product $C = AB$ is defined by

$$C = \begin{matrix} a\alpha+b\gamma & \alpha\beta+\beta\delta. \\ c\alpha+d\gamma & c\beta+d\beta \end{matrix}$$

Laguerre designs a system ('système simple') of the form

m	0	0	0
0	m	0	0
0	0	m	0
0	0	0	m

by the symbol m and he observes that $mA = Am$. He developed a calculus for the systems, for instance he studied equations of the type $AX = B$ and quadratic forms. On page 228 he remarks:

Soit A un système linéaire d'ordre n; je ne considérerai dans ce qui suit que des systèmes pouvant s'exprimer au moyen seulement de A et de systèmes simples. (Ceux-ci, en effet, se comportent comme de véritables nombres, en donnant ici le nom de *nombre*, indépendamment de toute idée arithmétique, aux quantités ordinaires, et réservant la dénomination de *quantités* aux systèmes linéaires proprement dits.)

It is Laguerre's aim to give a unification of several algebraic systems:

Le calcul des systèmes linéaires donne ainsi une interprétation simple et pour ainsi dire arithmétique des imaginaires, des quaternions, des clefs algébriques de Cauchy, des imaginaires congruentielles de Galois. (l.c. p. 235).

In the introduction I mentioned before Poincaré wrote (p. x):

Sans doute, il n'y a dans tout cela qu'une notation nouvelle, mais qu'on ne se trompe pas: dans les Sciences mathématiques, une bonne notation a la même importance philosophique qu'une bonne classification dans les Sciences naturelles.[1] (...) Depuis le commencement du siécle, de grands efforts ont été faits pour généraliser le concept de grandeur; des quantités réelles, on s'est élevé aux quantités imaginaires, aux nombres complexes, aux idéaux, aux quaternions, aux imaginaires de Galois. Le domaine de l'Analyse s'agrandissait ainsi sans cesse et de tous côtés; Laguerre s'élève à un point de vue d'où l'on peut embrasser d'un coup d'oeil tous les horizons. Toutes ces notions nouvelles, et en particulier les quaternions, sont ramenées aux substitutions linéaires.

In this work there are applications on the theory of abelian functions. All this may suffice to show that this work of Laguerre contains a germ of linear algebra.

In 1891 Carvallo published a long paper [15] on linear systems, referring to the 'systèmes linéaires' of Laguerre. He studied vector functions and gave several applications; that is, X and Y being vectors, he studied transformations $Y = \varphi(X)$. He calls such a transformation, transforming X into Y, an 'opérateur'. It is interesting to read his remarks on the distinction between an 'opérateur' and a linear system (i.e. a matrix) ([15], p. 179):

Pour bien faire comprendre la différence qui existe entre les notions d'opérateur et de système linéaire, il nous suffira de dire que, si on change le système des axes coordonnés, on obtient un système linéaire différent pour représenter la même fonction vectorielle et par suite de même opérateur.

With reference to this paper Peano published in 1894 a note in the same journal [97] in which he called Carvallo's notes 'de la plus grande importance'. Peano filled in an omission in Carvallo's paper

[1] Compare the following remark of Gauss. In the *Disquisitiones Arithmeticae* (Werke I, art. 76, p. 60; Recherches arith. p. 56 [39]) he writes, dealing with some number-theoretical problems: 'Sed neuter demonstrare potuit, et cel. Waring fatetur demonstrationem eo difficiliorem videri, quod nulla *notatio* fingi possit, quae numerum primum exprimat.—At nostro quidem iudicio huiusmodi veritates ex notionibus potius quam ex notationibus hauriri debebant.' ['Neither of them has been able to prove it and the famous Waring lets appear that to him the proof seems to be the more difficult because he cannot invent a notation for expressing a prime number. But to my opinion one must draw these sorts of statements from the concepts and not from the notations.']

with respect to the convergence of the series

$$e^\varphi = 1 + \ldots + \frac{\varphi^n}{n!} + \ldots,$$

φ being a linear system. Peano remarked that he already treated this question in a general form in his *Calcolo geometrico* (1888) [96], on which I shall comment later on in this chapter, where he generalizes 'des théorèmes de l'Analyse, généralisés aux nombres complexes d'ordre quelconque' ([97], p. 136).

§ 3 *Grassmann*

The next subject I have to discuss in the development of the concept of linear space is the work of Hermann Grassmann (1809–1877).

Grassmann's work is known in mathematics as *die Ausdehnungslehre von Grassmann*. The background of his work is the tendency towards direct, coordinate-free methods in geometry which I described above and, with a very global characterization, one may say that the Ausdehnungslehre contains a symbolic calculus with the geometric objects themselves, that is without using coordinates. This led him to propositions, which are very close to the concepts of the modern theory of linear spaces and linear algebra. How close this was will be seen from passages from his work, which I quote in detail. First I give the bibliography.

The first publication of the Ausdehnungslehre was in 1844. The book, published in Leipzig, was entitled: *Die Wissenschaft der extensiven Grösse oder die Ausdehnungslehre, Erster Theil, die lineale Ausdehnungslehre enthaltend*. The title of the 'Erster Theil' is: *Die lineale Ausdehnungslehre, ein neuer Zweig der Mathematik, dargestellt und durch Anwendungen auf die übrigen Zweige der Mathematik, wie auch auf die Statik, Mechanik, die Lehre vom Magnetismus und die Krystallonomie erläutert* [43].

Owing to the fact that this book was very hard to read,–it is indeed almost unreadable–Grassman prepared a new version of his theory, published in 1862: *Die Ausdehnungslehre, vollständig und in strenger Form bearbeitet*, now known as 'die Ausdehnungslehre von 1862' [44].

In 1847 Grassmann published *Geometrische Analyse, geknüpft an die von Leibniz erfundene geometrische Characteristik, gekrönte Preisschrift der Fürstlich Jablonowski'schen Gesellschaft zu Leipzig.* The Ausdehnungslehre from 1844 and the Geometrische Analyse from 1847 were edited in 1894 in *Hermann Grassmanns gesammelte mathematische und physikalische Werke; herausgegeben von Friedrich Engel; Ersten Bandes erster Theil: die Ausdehnungslehre von 1844 und die Geometrische Analyse.* The work from 1862 was edited in *Ersten Bandes zweiter Theil: Die Ausdehnungslehre von 1862, in gemeinschaft mit Hermann Grassmann dem jüngeren herausgegeben von Friedrich Engel,* Leipzig 1896 [45].

Grassmann's work was very original in his time. It is highly abstract in nature. His 1844-book bubbled over with philosophical considerations and even now it is difficult to read. But the edition from 1862 –in which the mathematical form is much better and which is quite readable–also remained rather unknown for a long time. In the preface to the edition from 1894 [(45] Vorbemerkungen p. v) Engel wrote:

... man begnügt sich im günstigsten Falle, seinen Namen mit einer gewissen Hochachtung zu nennen, aber an seinen Werken geht man achselzuckend vorüber.

Hankel seems to have been the first to see the importance of Grassmann's work; see [56], [45a], [48], [145]. In his *Theorie der complexen Zahlensysteme* Hankel studies formal mathematical systems ('formale Mathematik') and, in particular, formal combinations of several 'imaginäre Einheiten' (p. 99), satisfying adequate laws (addition, multiplication, associativity; commutativity of multiplication is not always supposed), leading to complex numbers, 'höheren complexen Zahlen', quaternions. He writes that Grassmann was the first to study such formal systems 'mit wahrhaft philosophischen Geiste ergriffen' (l.c. p. 16, 105). As to the fact that the considerable importance of Grassmann's work was not recognized, Hankel writes (l.c. p. 16):

Wenn trotzdem die Untersuchungen des geistreichen Forschers die Anerkenning nicht gefunden haben, die sie verdienen, und die jeder, der sie kennt, ihnen zollen muss, so ist dies, meines Erachtens, hauptsächlich dem Umstande zuzuschreiben, dass ihr Verfasser allen Sätzen sogleich die allgemeinsten Form (in Bezug auf *n* Dimensionen) gegeben, dadurch aber die Uebersichtlichkeit und das Verständniss ungemein erschwert hat.

Now about the work of Grassmann and its place in the development of the concept of linear space.

The theory of Grassmann consists of a kind of calculus or algebra for elements whose nature he does not specify, that is he defines addition and a certain kind of multiplication (which is a very important operation for the theory) for the elements in such a way that a hierarchy of structures is generated – Grassmann calls them 'Stufen. – The application to geometrical objects – points, line segments, planes – leads to a calculus for geometric objects, with certain analogies to operations in algebra.

In view of a comparison with the notions of 'Richtung' and 'entgegengesetzte Richtung' such as were introduced by Bolzano I quote a passage from the introduction to the edition from 1844 ([45a], p. 8):

Den ersten Anstoss gab mir die Betrachtung des Negativen in der Geometrie; ich gewöhnte mich, die Strecken AB und BA als entgegengesetzte Grössen aufzufassen; woraus denn hervorging, dass, wenn A, B, C, Punkte einer geraden Linie sind, dann auch allemal $AB + BC = AC$, sei, sowohl wenn AB und BC gleichbezeichnet sind, als auch wenn entgegengesetzt bezeichnet, dass heisst wenn C zwischen A und B liegt.

It seems that the origin of Grassmann's theory is to be found in the classical book of Möbius (1790–1868) *Der baryzentrische Calcul* which was published in 1827, and in particular in the following theorem of Möbius ([87], p. 31):

Ist eine beliebige Anzahl $= v$ Puncten A, B, C, ..., N mit resp. Coefficienten a, b, c, ..., n gegeben, deren Summe nicht $= 0$ ist, so kann immer ein Punct S, und nur einer, – der Schwerpunct, – von der Beschaffenheit gefunden werden, dass, wenn man durch die gegebenen Puncte und den Punct S nach einer beliebigen Richtung Parallelen zieht, und diese mit einer willkürlich gelegten Ebene schneidet, welches resp. in A', B', C', ..., N', S' geschehe, dass dann immer

$$a \cdot AA' + b \cdot BB' + c \cdot CC' + ... + n \cdot NN' = (a + b + c + ... + n) \cdot SS'$$

und folglich, wenn die Ebene durch S selbst geht

$$a \cdot AA' + b \cdot BB' + c \cdot CC' + ... + n \cdot NN' = 0.$$

For the relation between $A, B, C, ..., N, S$ Möbius introduced the notation

$$aA + bB + cC + ... + nN = (a + b + c + ... + n)S;$$

this is the addition of points.

In the review paper [48] on Grassmann's life and work one finds (p. 6):

> Diese Schreibweise und die durch sie definierte Addition von Punkten darf vielleicht neben gewissen mechanischen Betrachtungen als der Quellpunkt der Grassmann'schen Ausdehnungslehre angesehen werden; Grassmann geht aber von einer anderen Construction des Punktes S aus, wobei er die geometrische Addition von Strecken benutzt.

Grassmann considers a system of elements on which he defines in a formal way the operations of addition and multiplication called 'Verknüpfungen'. He calls the elements *extensive Grössen*. The objects of geometry, for instance, are extensive Grössen.[1]

In a later paper (1874) [47], in which he returned to the Ausdehnungslehre, Grassmann gave the following explanation, which is preceded by the remark:

> Die neuere Algebra hat durch die vereinten Bemühungen der hervorragendsten Mathematiker gegenwärtig eine Ausbildung erlangt, welche sie fast mit allen Zweigen der Mathematik in die engste Beziehung setzt und auch diese mit ihren Ideen befruchtet.[2] (...)
> Den extensiven Grössen, welche die Ausdehnungslehre behandelt, liegt eine Reihe von Grössen zu Grunde, welche in keiner Zahlbeziehung zu einander stehen, d.h. von denen sich keine aus den übrigen numerisch ableiten oder, anders ausgedrückt, keine sich als lineare Funktion der übrigen mit Zahlcoefficienten darstellen läszt,

---

[1] Grassmann introduced the denomination 'extensive Grösse' to distinguish the objects of geometry from the objects of the theory of functions, which he called 'intensive Grössen'. I omit the details of his motivation–which is rather philosophical–; Grassmann remarks that geometry is concerned with generation of objects by different kinds of 'Einheiten', while in function theory one is concerned with the generation by means of the same kind of 'Einheit'.

[2] Grassmann mentions that Clebsch inspired him to this new study of the Ausdehnungslehre.

und die ich, sofern sie als ursprünglich zu Grunde liegene betrachtet werden, Einheiten erster Stufe genannt habe. Als solche können z.B. im Raum 4 beliebige Punkte betrachtet werden, die nicht in einer Ebene liegen. Es seien $e_1, ..., e_m$ diese Einheiten, so nenne ich Grösse erster Stufe jede Grösse $x_1 e_1 + ... + x_m e_m$, wo $x_1, ..., x_m$ Zahlgrössen sind, und die Gesammtheit dieser Grössen nenne ich ein Gebiet m^{ter} Stufe. (...)

Das Product zweier Grössen erster Stufe nenne ich ein combinatorisches, wenn für dasselbe die Gesetze $(aa) = 0$, $(ab) = -(ba)$ gelten, und nenne diese Producte und die aus ihnen numerisch ableitbaren Grössen Grössen zweiter Stufe. Entsprechend bei 3 und mehr Factoren erster Stufe, bei denen gleichfalls das Product Null wird, wenn zwei Factoren gleich werden, und entgegengesetzten Werth annimmt, wenn man zwei desselben vertauscht. (l.c. p. 538). As to the product I remind the reader of the outer product of two vectors. As to the example of 4 points note that this is exactly the system which led Möbius to the introduction of the barycentric coordinates (homogeneous coordinates).

Nowadays we call the system Grassmann considered a *Grassmann algebra*; this is not the place to give the exact definition.[1]

Die Ausdehnungslehre von 1862 is preceded by a long introduction in which Grassmann not only gives a summary of his theory but which also contains a justification of his work and even a defence against the objection that Grassmann is concerned with 'Grössen' whereas the objects of his theory are not at all 'Grössen'. This objection is evidently an objection against the formal methods which are essential for Grassmann's theory. Grassmann's defence shows that he was quite close to the modern axiomatic methods in his ideas and that in this he was in great advance of the mathematicians of his time.

I quote some passages from the introduction at large because Grassmann's books are rather unknown and a simple statement that Grassmann was a predecessor of linear and multilinear algebra is not sufficient for convincing the reader–at least it ought not to be (l.c. p. 5–7):

---

[1] See for instance Bourbaki, *Algèbre multilinéaire* [12].

Die hier gewählte[1] schliesst sich am engsten an die Arithmetik an, doch in der Weise, dass sie die Zahlgrösse schon als eine stetige voraussetzt. Wie nun die Arithmetik alle übrigen Grössen aus einer einzigen, im Uebrigen willkürlichen Grösse, die als Einheit gesetzt wird, und mit e bezeichnet sein mag, entwickelt (vergleiche mein Lehrbuch der Arithmetik, 1861 Berlin bei Enslin), so setzt die Ausdehnungslehre in der hier gegebenen Fassung mehrere solche Grössen, e_1, e_2, ..., von denen keine aus den übrigen ableitbar ist, zum Beispiel e_2 sich nicht aus e_1 dadurch entwickeln lässt, dass e_1 mit irgend einer Zahlgrösse multipliert wird, voraus, und betrachtet zunächst die aus jenen Einheiten durch Multiplikation mit Zahlgrössen und Addition dieser Produkte entstandenen Grössen, welche ich extensive Grössen (oder Ausdehnungsgrössen) genannt habe. Hieraus ergeben sich denn leicht die in Kap. 1 vorgetragenen Gesetze der Addition, Subtraktion, Vielfachung (Multiplikation mit Zahlen) und Theilung (Division durch Zahlen).

Es mag auffallend erscheinen, dass diese so einfache Idee, welche im Grunde genommen in weiter nichts besteht, als dass eine Vielfachensumme verschiedener Grössen (als welche hiernach die extensive Grösse erscheint) als selbständige Grösse behandelt wird, in der That zu einer neuen Wissenschaft sich entfalten soll; und man hat mir denn auch, hieran anknüpfend, den Einwurf gemacht, dass die ganze Ausdehnungslehre nur eine abgekürzte Schreibart sei, ja, dass es fehlerhaft sei, Ausdrücke als Grössen zu behandeln, welche gar keine Grössen seien. Allein dieser Einwurf beruht auf einem gänzlichen Verkennen des Wesens der Mathematik und der Grössen. Auf diese Weise würde die ganze Arithmetik, ja, man kann sagen, die ganze reine Mathematik, bloss eine abgekürzte Schreibart sein; denn die Zahl ist nur ein abgekürzter Ausdruck für eine Summe von Einheiten, das Produkt für eine Summe gleicher Zahlen, die Potenz für ein Produkt solcher, und so weiter; dennoch würde ohne diese abgekürzte Schreibart, oder, um es richtiger auszudrücken, ohne diese Zusammenfassung zu einer Einheit des Begriffes kein Fortschritt denkbar sein. Es würde zum Beispiel ohne diese Zusammen-

---

[1] Grassmann refers to the method of composition of the new book in comparison to the edition from 1844.

fassung nicht möglich sein, zu dem Begriffe der wegnehmenden Rechnungsarten (Subtrahiren, Dividiren, Radiciren, Logarithmiren), und zu den durch sie neu sich entwickelnden Zahlformen: der negativen, gebrochenen, irrationalen und imaginären, zu gelangen. Es kommt überall nur darauf an, dass man auch wirklich dasjenige zusammenfasse, was seinem Wesen nach eine Einheit bildet, und was daher auch zu neuen Resultaten führen muss, zu denen man ohne jene Zusammenfassung nicht gelangen würde.

Die Ausdehnungslehre führt nun in der That zu einem unerschöpflichen Reichthum solcher Beziehungen, welche ohne Bildung jener Begriffseinheit, welche in der extensiven Grösse erscheint, auf keine Weise aufzufassen oder abzuleiten wären. Ob man diesem Begriffe den Namen einer Grösse zugesteht, ist an und für sich von sehr untergeordneter Bedeutung, da es hier auf Namen wenig ankommt. Die Frage ist nur die, ob dieser neue Begriff mit dem allgemeinen Begriffe der Grösse wirklich so zusammenhänge, dass sie ihrem Wesen nach zu einem Gesammtbegriffe sich zusammenschliessen, und dass eine zwischen beiden Gebieten gezogene Gränzlinie das Zusammengehörige willkürlich und der Sache widersprechend zertrennen würde. Ist letzteres der Fall, so wäre es sogar fehlerhaft, diesem neuen Begriffe nicht den Namen der Grösse beizulegen.

Nun glaube ich in der That, dass zwischen dem, was ich extensive Grösse genannt habe, und zwischen allgemeinen Zahlgrössen und namentlich der imaginären Grösse $(a+bi)$ eine so innige Beziehung herrscht, dass es widersinnig wäre, die eine als Grösse zu betrachten und die andere nicht, da ja in der That die imaginäre Grösse ebenso aus zwei Einheiten 1 und $i = \sqrt{-1}$ durch reelle Zahlkoefficienten ableitbar ist, wie die extensiven Grössen aus zwei oder mehr Einheiten ableitbar sind (...). So scheint es mir also vollständig gerechtfertigt, wenn ich die extensive Grösse als Grösse bezeichne. Aber ich gehe noch weiter, indem ich sie nicht nur als Grösse überhaupt, sondern auch als einfache Grösse bezeichne. Ihr treten nämlich gegenüber andere Grössen, welche den Charakter zusammengesetzter Grössen ebenso entschieden an sich tragen, wie jene den der einfachen, und welche erst durch Addition höherer Gebilde und besonders durch die Betrachtung der Quotienten und der Funktionen hineintreten (...).

Ich fahre nun fort, den Gang der Entwickelung in dem vorliegenden Werke übersichtlich zu verfolgen.

An die Addition, Subtraktion, Vielfachung und Theilung schliesst sich (in Kap. 2) der allgemeine Begriff der Multiplikation extensiver Grössen an, welcher auf die Beziehung der Multiplikation zur Addition (nämlich darauf, dass man statt der Summe die Summanden multipliciren darf) gegründet ist. Hiernach führt die Multiplikation der genannten Grössen auf die ihrer Einheiten ($e_1, e_2, \ldots$) zurück, und aus der Betrachtung der Produkte dieser Einheiten ergeben sich dann verschiedene Gattungen der Multiplikation. Es gelingt nun, aus diesen Gattungen zwei auszusondern, auf welche sich alle übrigen zurückführen lassen.

Die eine derselben fällt in ihren Gesetzen ganz zusammen mit der gewöhnlichen Multiplikation in der Algebra und ist daher von mir die algebraische genannt worden. Aber sie ist in Bezug auf die durch sie erzeugten Grössen bei weitem die verwickeltste und kann nur durch Betrachtung der Funktionen zur vollen Klarheit gebracht werden, weshalb ich sie auf den zweiten Abschnitt dieses Werkes verwiesen habe. Die Bezeichnung für die algebraische Multiplikation muss der Natur der Sache nach mit der gewöhnlichen Bezeichnung der Multiplikation zusammenfallen, da es widersinnig wäre, Verknüpfungen, welche in allen Beziehungen denselben Gesetzen unterliegen, verschieden zu bezeichnen.

Die zweite jener Multiplikationen, welche im dritten Kapitel behandelt ist, zeigt sich als die für die Ausdehnungslehre charakteristische, und sie wesentlich weiter fördernde, indem sie die verschiedenen Stufen einfacher Grössen liefert, welche in der Ausdehnungslehre hervortreten. Sie ist dadurch gekennzeichnet, dass zwei einfache Faktoren des Produktes nur vertauscht werden dürfen, wenn man zugleich das Vorzeichen ($+ \; -$) des Produktes umkehrt. Da zwar für diese Multiplikation die Beziehung zur Addition dieselbe ist, wie bei jeder Multiplikation, aber die übrigen Gesetze derselben wesentlich von denen der gewöhnlichen Multiplikation abweichen, so war es nothwendig, sie durch die Bezeichnung zu unterscheiden. Ich habe in diesem Werke dafür die Bezeichnung durch eckige Klammern, die das Produkt umschliessen, gewählt, so dass also $[ab] = -[ba]$ ist, wenn a und b einfache Faktoren dieses Produktes sind. Es entfaltet sich dies Produkt zu einer ausserordentlichen Mannigfaltigkeit von Erscheinungsformen, und lässt in reicher Fülle Beziehungen hervortreten, welche auf alle Zweige der Mathematik ein unerwartet neues Licht werfen, so dass es den eigentlichen Mittelpunkt der neuen Wissenschaft bildet.[1]

I omit Grassmann's summary of the 'Zweiter Abschnitt' which is not relevant for our purpose.

I quote some pages from the 'Erster Abschnitt'; the reader may compare them with the concepts of modern linear algebra. Here Grassmann deals with some basic concepts of his theory: the addition and scalar multiplication of extensive Grössen. His definitions are strictly formal without philosophical motivation.

Erster Abschnitt

Die einfachen Verknüpfungen extensiever Grössen

Kapitel 1. Addition, Subtraktion, Vervielfachung und Theilung extensiver Grössen

§ 1. Begriffe und Rechnungsgesetze

1. Erklärung. Ich sage, eine Grösse a sei aus den Grössen $b, c, \ldots$ durch die Zahlen $\beta, \gamma, \ldots$ abgeleitet, wenn

$$a = \beta b + \gamma c + \ldots$$

ist, wo $\beta, \gamma, \ldots$ reelle Zahlen sind, gleichviel ob rational oder irrational, ob gleich Null oder verschieden von Null. Auch sage ich, a sei in diesem Falle numerisch abgeleitet aus $b, c, \ldots$.

---

1 The introduction of an anticommutative product reminds one of the theory of Lie algebras. In various places in the works of Lie there are references to Grassmann. In a footnote in his paper [82] he writes about 'der modernen Mannigfaltigkeitslehre, deren Ursprung wohl auf Grassmann zurückzuführen ist'. See also *Theorie der Transformationsgruppen* III where Lie mentions Grassmann's complex numbers in the introduction ([83], p. XXII); there are also some remarks on the relation between Grassmann's work from 1844 and the methods which Riemann and Lie himself used in the foundation of geometry. See also [83] p. 747 on hypercomplex numbers.

2. Erklärung. Ferner sage ich, dass zwei oder mehrere Grössen $a, b, c, \ldots$ in einer Zahlbeziehung zu einander stehen, oder dass der Verein der Grössen $a, b, c, \ldots$ einer Zahlbeziehung unterliege, wenn irgend eine derselben sich aus den übrigen numerisch ableiten lässt, also wenn sich zum Beispiel

$$a = \beta b + \gamma c + \ldots$$

setzen lässt, wo $\beta, \gamma, \ldots$ reelle Zahlen sind. Besteht der Verein nur aus Einer Grösse a, so soll nur in dem Falle gesagt werden, der Verein unterliege einer Zahlbeziehung, wenn $a = 0$ ist.

Wenn *zwei* Grössen a und b, von denen keine null ist, in einer Zahlbeziehung zu einander stehen, so bezeichne ich dies durch

$$a \equiv b,$$

und sage a sei kongruent b.

Anmerkung. Zwei reelle Zahlen stehen also immer, zwei verschieden benannte Grössen stehen nie in einer Zahlbeziehung zu einander. Null ist aus jeder Grössenreihe numerisch ableitbar, nämlich durch die Zahlen $0, 0, \ldots$. Mehrere Grössen also, unter denen eine null ist, stehen stets in einer Zahlbeziehung zu einander.

Das Zeichen ($\equiv$) ist in ähnlichem Sinne von Möbius (in seinem barycentrischen Kalkül) gebraucht. Die Benennung (kongruent) gründet sich auf geometrische Betrachtungen. Zur Bezeichnung abstrakter Beziehungen ist sie von Gauss gebraucht.

3. Erklärung. Einheit nenne ich jede Grösse, welche dazu dienen soll, um aus ihr eine Reihe von Grössen numerisch abzuleiten, und zwar nenne ich die Einheit eine ursprüngliche, wenn sie nicht aus einer anderen Einheit abgeleitet ist. Die Einheit der Zahlen, also die Eins, nenne ich die absolute Einheit, alle übrigen relative. Null soll nie als Einheit gelten.

4. Erklärung. Ein System von Einheiten nenne ich jeden Verein von Grössen, welche in keiner Zahlbeziehung zu einander stehen, und welche dazu dienen sollen, um aus ihnen durch beliebige Zahlen andere Grössen abzuleiten.

Anm. Hierher gehört auch der Fall, wo der Verein nur aus einer Einheit besteht (die jedoch nach Nr. 3 nicht null sein darf).

5. Erklärung. Extensive Grösse nenne ich jeden Ausdruck, wel-

cher aus einem Systeme von Einheiten (welches sich jedoch nicht auf die absolute Einheit beschränkt) durch Zahlen abgeleitet ist, und zwar nenne ich diese Zahlen die zu den Einheiten gehörigen Ableitungszahlen jener Grösse; zum Beispiel ist das Polynom

$$\alpha_1 e_1 + \alpha_2 e_2 + \dots,$$

oder

$$\sum \alpha e \text{ oder } \sum \alpha_r e_r,$$

wenn $\alpha_1, \alpha_2, \dots$ reelle Zahlen sind, und $e_1, e_2, \dots$ ein System von Einheiten bilden, eine extensive Grösse, und zwar ist dieselbe aus den Einheiten $e_1, e_2, \dots$ durch die zugehörigen Zahlen $\alpha_1, \alpha_2, \dots$ abgeleitet. Nur wenn das System bloss aus der absoluten Einheit (1) besteht, ist die abgeleitete Grösse keine extensive, sondern eine Zahlgrösse.

Den Ausdruck *Grösse* überhaupt werde ich nur für diese beiden Gattungen derselben festhalten. Wenn die extensive Grösse aus den *ursprünglichen Einheiten* abgeleitet werden kann, so nenne ich jene Grösse eine extensive Grösse erster Stufe.

Anm. Aus der Elementarmathematik setzen wir die Rechnungsgesetze für Zahlen, und auch für die sogenannten 'benannten Zahlen', das heisst, für die aus Einer Einheit abgeleiteten extensiven Grössen voraus; jedoch nur für den Fall, dass jene Einheit eine ursprüngliche ist.

6. Erklärung. Zwei extensive Grössen, die aus demselben System von Einheiten abgeleitet sind, addiren, heisst, ihre zu denselben Einheiten gehörigen Ableitungszahlen addiren, das heisst,

$$\sum \alpha e + \sum \beta e = \sum (\alpha + \beta) e.$$

7. Erklärung. Eine extensive Grösse von einer andern, aus demselben Systeme von Einheiten abgeleiteten subtrahiren, heisst die Ableitungszahlen der ersteren von den zu denselben Einheiten gehörigen Ableitungszahlen der letzteren subtrahiren, das heisst,

$$\sum \alpha e - \sum \beta e = \sum (\alpha - \beta) e.$$

Anm. In Bezug auf die Klammerbezeichnung halte ich die Bestimmung fest, dass ein ohne Klammern geschriebenes Polynom oder Produkt aus mehreren Faktoren gleichbedeutend ist dem mit

Klammern geschriebenen Ausdruck, in welchem alle Klammern gleich zu Anfang eintreten, also

$$a + b + c = (a + b) + c, \quad abc = (ab)c$$

und so weiter.

8. *Für extensive Grössen a, b, c gelten die Fundamentalformeln*:

1) $$a + b = b + a,$$

2) $$a + (b + c) = a + b + c,$$

3) $$a + b - b = a,$$

4) $$a - b + b = a.$$

Beweis. Es sei

$$a = \sum \alpha e, \quad b = \sum \beta e, \quad c = \sum \gamma e,$$

so ist

1) $$\begin{aligned}
a + b &= \sum \alpha e + \sum \beta e = \sum (\alpha + \beta)e \qquad \text{[nach 6]} \\
&= \sum (\beta + \alpha)e = \sum \beta e + \sum \alpha e \qquad \text{[6]} \\
&= b + a.
\end{aligned}$$

2) $$\begin{aligned}
a + (b + c) &= \sum \alpha e + (\sum \beta e + \sum \gamma e) \\
&= \sum \alpha e + \sum (\beta + \gamma)e \qquad \text{[6]} \\
&= \sum (\alpha + (\beta + \gamma))e \qquad \text{[6]} \\
&= \sum (\alpha + \beta + \gamma)e \\
&= \sum (\alpha + \beta)e + \sum \gamma e \qquad \text{[6]} \\
&= \sum \alpha e + \sum \beta e + \sum \gamma e \qquad \text{[6]} \\
&= a + b + c.
\end{aligned}$$

3) $$\begin{aligned}
a + b - b &= \sum \alpha e + \sum \beta e - \sum \beta e \\
&= \sum (\alpha + \beta)e - \sum \beta e \qquad \text{[6]} \\
&= \sum (\alpha + \beta - \beta)e \qquad \text{[7]} \\
&= \sum \alpha e = a.
\end{aligned}$$

4) $$\begin{aligned}
a - b + b &= \sum \alpha e - \sum \beta e + \sum \beta e \\
&= \sum (\alpha - \beta)e + \sum \beta e \qquad \text{[7]} \\
&= \sum (\alpha - \beta + \beta)e \qquad \text{[6]} \\
&= \sum \alpha e = a.
\end{aligned}$$

For this quotation see [45b], p. 11–13.

Grassmann summarizes these results in the following theorem ([45b], p. 14): 'Für extensive Grössen gelten die sämtlichen Gesetze algebraischer Addition und Subtraktion.'

Next Grassmann deals with scalar multiplication, that is the multiplication by real numbers. He defines

$$\sum \alpha e \cdot \beta = \beta \cdot \sum \alpha e = \sum (\alpha \beta) e.$$

The following properties, in which a, b are 'extensive Grössen' and β, γ real numbers, are proved:

$$a\beta = \beta a,$$
$$a\beta\gamma = a(\beta\gamma),$$
$$(a+b)\gamma = a\gamma + b\gamma,$$
$$a(\beta+\gamma) = a\beta + a\gamma,$$
$$a \cdot 1 = a,$$

$a\beta = 0$ implies $a = 0$ or $\beta = 0$.

They are summarized in the theorem ([45b] p. 16):

Für die Multiplikation und Division extensiver Grössen durch Zahlen gelten die algebraische Gesetze der Multiplikation und Division.

One recognizes the familiar laws of linear algebra. But Grassmann obtains his results for formally defined sums of elements (Grössen) and as a consequence associativity and distributivity appear with Grassmann as properties that ought to be proved in contradiction to our axiomatic theory, where these properties are postulated. One may doubt whether Grassmann had in mind something like a kind of space, a vector space, as we do. A concept like that seems to have been reserved for later years.

Grassmann developed the theory of the Verknüpfungen; he obtained results to be compared with our concepts of linearly dependent and linearly independent sets of elements. Again I quote some pages ([45b], p. 16, 17):

§ 2. Zusammenhang zwischen den aus einem System von Einheiten ableitbaren Grössen

14. Erklärung. Die Gesammtheit der Grössen, welche aus einer Reihe von Grössen $a_1, a_2, \ldots, a_n$ numerisch ableitbar sind, nenne ich das aus jenen Grössen ableitbare Gebiet (das Gebiet der Grössen $a_1, \ldots, a_n$), und zwar nenne ich es ein Gebiet n-ter Stufe, wenn jene Grössen von erster Stufe (das heisst, aus n ursprünglichen Einheiten numerisch ableitbar) sind, *und sich das Gebiet nicht aus weniger als n solchen Grössen ableiten lässt.* Ein Gebiet, welches ausser der Null keine Grösse enthält, heisst ein Gebiet nullter Stufe.

Anm. Das Gebiet erster Stufe ist also die Gesammtheit der Vielfachen einer Grösse erster Stufe, wenn man nämlich unter *Vielfachem einer Grösse* jedes Produkt der Grösse mit einer reellen Zahlgrösse versteht.

15. Erklärung. Zwei Gebiete heissen identisch, wenn jede Grösse des ersten Gebietes zugleich Grösse des zweiten ist, und umgekehrt. Wenn jede Grösse eines Gebietes (*A*) zugleich Grösse eines andern (*B*) ist (ohne dass das Umgekehrte nothwendig stattfindet), so nenne ich beide Gebiete einander incident, und sage dann, das erste Gebiet (*A*) sei dem zweiten untergeordnet, das zweite dem ersten übergeordnet. Die Gesammtheit der Grössen, welche zweien oder mehreren Gebieten zugleich angehören, heisst ihr gemeinschaftliches Gebiet, und die Gesammtheit der Grössen, welche sich aus den Grössen zweier oder mehrerer Gebiete numerisch ableiten lassen, ihr verbindendes Gebiet.

Anm. Ist zum Beispiel das Gebiet A aus den Einheiten e_1, e_2, e_3 abgeleitet und das Gebiet B aus den Einheiten e_2, e_3, e_4, so ist das den Gebieten A und B gemeinschaftliche Gebiet das aus den Einheiten e_2, e_3 abgeleitete, und das A und B verbindende Gebiet das aus den Einheiten e_1, e_2, e_3, e_4 abgeleitete.

16. *Zwischen n Grössen $a_1, \ldots, a_n$ herrscht dann und nur dann eine Zahlbeziehung, wenn sich eine Gleichung*

$$\alpha_1 a_1 + \ldots + \alpha_n a_n = 0$$

aufstellen lässt, in welcher die Zahlen $\alpha_1, \ldots, \alpha_n$ nicht alle zugleich null sind.

Beweis. Denn, wenn in der Gleichung

$$\alpha_1 a_1 + \ldots + \alpha_n a_n = 0$$

auch nur Eine der Zahlen $\alpha_1, \ldots, \alpha_n$ von Null verschieden ist, zum Beispiel α_1, so ist die mit dieser Zahl verbundene Grösse a_1 aus den übrigen numerisch ableitbar; denn dann ist

$$a_1 = -\frac{\alpha_2}{\alpha_1} a_2 - \frac{\alpha_3}{\alpha_1} a_3 - \ldots - \frac{\alpha_n}{\alpha_1} a_n.$$

Umgekehrt, wenn irgend eine Zahlbeziehung zwischen den Grössen $a_1, \ldots, a_n$ herrscht, zum Beispiel

$$a_1 = \beta_2 a_2 + \beta_3 a_3 + \ldots + \beta_n a_n,$$

so wird

$$-a_1 + \beta_2 a_2 + \beta_3 a_3 + \ldots + \beta_n a_n = 0,$$

eine Gleichung, in welcher wenigstens der Koefficient von a_1 ungleich Null ist.

17. *Wenn n Grössen in einer Zahlbeziehung zu einander stehen, und sie nicht alle null sind, so muss sich aus ihnen ein Verein von weniger als n Grössen aussondern lassen, welcher keiner Zahlbeziehung unterliegt, und aus dem die übrigen Grössen numerisch ableitbar sind.*

I omit his proof of this theorem. Grassmann knew our concept of dimension, although he did not use this term. He proved the following theorem ([45b], p. 21), in which the Stufenzahl corresponds to our concept of dimension:

Die Stufenzahlen zweier Gebiete sind zusammengenommen ebenso gross als die Stufenzahlen ihres gemeinschaftlichen und ihres verbindenden Gebietes zusammengenommen, dasz heiszt, wenn m und n die Stufenzahlen der gegebenen Gebiete sind, r die ihres gemeinschaftlichen, v die ihres verbindenden Gebieten, so ist

$$m + n = r + v.$$

Until here it concerns the theory of linear spaces. But Grassmann defines also a multiplication of extensive Grössen (see my quotations from the introduction). The multiplication of $a = \sum \alpha_i e_i$ and $b = \sum \beta_j e_j$

is reduced to the definition of the products $e_i e_j$. The relations between these products are called the 'Bestimmungsgleichungen'. Grassmann considered the case that $e_i e_j = -e_j e_i$ for $i \neq j$ and so $e_i^2 = 0$.[1]

In this way one obtains–in our terminology–an algebra. I don't pursue the subject, because my aim is the survey of the development of the concept of linear space and not of linear (multilinear) algebras.

As to the objection against the 'extensive Grössen' I mentioned above it is interesting to read Möbius's comment in his paper 'Die Grassmann'sche Lehre von Punctgrössen und den davon abhängenden Grössenformen', see Möbius' *Werke* ([87], p. 615):

Das Studium der voranstehenden Abhandlung des Herrn Grassmann und besonders des letzteren Theils derselben dürfte, ungeachtet des nicht zu verkennenden Strebens ihres Verfassers nach Klarheit, dennoch mit einigen Schwierigkeiten verknüpft sein, welche daraus hervorgehen, dass der Verfasser seine neue geometrische Analysis auf eine Weise zu begründen sucht, welche dem bisher bei mathematischen Betrachtungen gewohnten Gange ziemlich fern liegt, und dass er nach Analogien mit arithmetischen Operationen Objecte als Grössen behandelt, die an sich keine Grössen sind, und von denen man sich zum Theil keine Vorstellung bilden kann. Da gleichwohl diese neue Analysis wegen der Einfachheit, mit welcher sich geometrische Untersuchungen durch sie führen lassen, alle Aufmerksamkeit zu verdienen scheint, so habe ich es im Folgenden versucht, sie auf eine dem Geiste der Geometrie entsprechendere und damit, wie ich hoffe, leichter fassliche Weise zu begründen und zu zeigen, wie jene Scheingrössen als abgekürzte Ausdrücke wirklicher Grössen angesehen werden können. Uebrigens habe ich meinen Aufsatz nicht in der Form eines Commentars, sondern als eine selbständige Abhandlung abgefasst, so dass er Jedem auch noch vor Lesung des Vorhergehenden verständlich sein wird.–Die von Herrn Grassmann eingeführten Benennungen habe ich unverändert beibehalten, sowie ich auch bei rein mathematischen Entwickelun-

---

1 Grassmann used the term 'lineale Multiplikation' (compare the title in the edition from 1844, Erster Abschnitt). 'Lineal nennt er eine solche Productbildung, deren Bestimmungsgleichungen von den Einheiten auf die daraus beliebig linear zusammengesetzten extensiven Grössen sich übertragen.' ([48], p. 10). The condition $ab = ba$ leads to 'ordinary' algebra.

gen dieses Geometers von seinem Gange nicht wesentlich abge-
wichen bin. Was sonst noch von mir hinzugefügt worden, ist nicht
von dem Belange, um hier besonders darauf aufmerksam zu machen.
Grassmann gave several applications of his theory, for instance in
geometry and mechanics. I quote from his paper 'Die Mechanik nach
den Principien der Ausdehnungslehre' [46], p. 222):

Es gibt wohl kaum ein Gebiet, auf welchem sich die Unentbehrlich-
keit der in meiner Ausdehnungslehre (von 1844 und 1862) darge-
stellten Kalküls so schlagend erwiesen wie in die Mechanik.

Finally a question of priority between Grassmann and Cauchy. In the
introduction to the Ausdehnungslehre von 1862 ([45b], p. 9) Grassmann observes that in 1845 Saint-Venant had published some work
on the geometric multiplication of line-segments, which agrees with
his outer product. He remarks that Saint-Venant evidently did not
know Grassmann's work from 1844 and that he had sent two items
of his book to Cauchy, asking Cauchy to send one to Saint-Venant.
In 1853 Cauchy published a paper in the *Comptes Rendus de l'Académie des Sciences* ([16]; see also [17]) in which he introduced a symbolic
method ('les clefs algébriques'), useful in solving certain algebraic
problems, that agrees exactly with Grassmann's method. Then Grassmann wrote (l.c. p. 9): 'Ich bin weit davon entfernt, den berühmten
Mathematiker eines Plagiats beschuldigen zu wollen...', but nevertheless he found it necessary to address a 'Prioritätsreklamation' to
the Académie in Paris. This was put in charge of a committee in 1854
(*C.R.* t. 38, p. 743), but there never was a reaction from this Committee (see [48], p. 11).[1]

Grassmann was a most remarkable scholar. The lack of recognition
of the merits of his mathematical work led him to the study of lin-
guistics, sanskrit, gothic and some other subjects. In this domain he
found recognition; see [48], p. 3 and [10].

For a detailed exposition on the influence of Grassmann's work on
the development of vector analysis in the classical sense see [19].

---

1 For more information about this priority question see ([19], p. 82).

§ 4 *Peano*

As I observed Grassmann's work had only little influence on his contemporary mathematicians. The next mathematician to play a role in the development in the concept of a linear space was *Peano*.

In 1888 Peano published a book *Calcolo geometrico secondo l'Ausdehnungslehre di H. Grassmann preceduto dalle operazioni della logica deduttiva* [96]. In this book one finds the axiomatic definition of a linear space, almost in our present formulation. But it is most remarkable that this book too had only little influence on the development of mathematics.

It was the aim of the author to develop a geometrical calculus in this book which he described as follows in the first chapter ('Formazioni geometriche'):

Il calcolo geometrico consiste in un sistema di operazioni analoghe a quelle del calcolo algebrico, ma in cui gli enti sui quali si eseguiscono i calcoli, inveci che numeri, sono enti geometrici, che definiremo.[1] (l.c. p. 21).

He observes that a first attempt for such a calculus was made by Leibniz (*Math. Schriften*, Berlin 1849, t. II, p. 17, t. V, p. 133). He mentions the work of Möbius [87] from 1827, some work of Bellavitis (1832)–for which he refers to Laisant, *Théorie et applications des équipollences*, Paris 1887[2]–, the work of Hamilton on the theory of quaternions (1853) and finally he refers to Grassmann's Ausdehnungslehre from 1844. He does not mention the Ausdehnungslehre from 1862.[3]

The most interesting parts of the book are the introduction ('Operazioni della logica deduttiva') and chapter IX ('Trasformazioni di sistemi lineari').

1 'The geometrical calculus consists of a system of operations analogous to those of algebraic calculus but in which the objects with which the calculations are performed are, instead of numbers, geometrical objects which we shall define.'

2 Bellavitis created the calculus of equipollences. A line-segment is designed by two letters; *AB* is considered to be different from *BA*. Two straight lines are called equipollent if they are equal, parallel and directed in the same sense. The concept of equipollent-sum is defined and a calculus of equipollences is developed. This is also a kind of vector-theory. For more information see [19].

3 For the history of the work of Möbius and the theory of quaternions see [19].

The introductory chapter contains the theory of logical operations about which Peano observes (p. VII): 'esse presentano grande analogia con quelle dell'algebra, e del calcolo geometrico.'[1] That is to say Peano defines the set-theoretical operations, and gives their properties; he states the relations between these set-theoretical operations and the logical operations (conjunction, disjunction, negation). It is the set-theory that is interesting because Peano introduced several of our modern notations, which, however, were not accepted by the mathematicians in those days; they appear in mathematics again after the second world war (independent of Peano?) and since then they are common notations. I give some examples.

Peano considers a system of elements ('enti') and subsets $A, B, \ldots$ of this system ('classi di questo sistema'), but sometimes he also used the terminology 'insieme' (set). He defines the equality of the classes A and B (every element of A belongs to B and conversely; notation $=$); a definition of the intersection of the classes A and B is given (maximal class contained in A and B). He denotes the intersection by $A \cap B$, sometimes also by AB. The union of A and B is defined (the minimal class which contains A and B) and denoted by $A \cup B$. The empty set is introduced for which Peano used the symbol $\bigcirc$. The 'universe', i.e. the basic set for which a theory is developed, is denoted by ●. Then Peano gives the usual rules for a calculus of sets: associativity, distributivity and so on.

Remarks. One may assume that this is the first time in the theory of sets that the symbols $\cap$ and $\cup$ appear for the intersection resp. the union. I remark that Grassmann in his Ausdehnungslehre from 1844 used the symbols $\smile$ and $\frown$ for the abstract operations ('Verknüpfungen') which he defines in his formal systems, but in the work from 1862 they don't appear anymore. Did Peano who referred to Grassmann-1844, take these symbols from Grassmann?

Peano mentions E. Schröder (*Der Operationskreis des Logikkalkuls*, Leipzig 1867), who used the notation $\times$ and $+$ in logic, and he states that he prefers $\cap$ and $\cup$ in set-theory just to avoid confusion with ordinary arithmetic. The symbols $\cap$ and $\cup$ were not accepted by the

---

1 'it presents great analogy with those of algebra and of geometrical calculus.'

mathematicians for many years. In the older literature on set-theory and topology one finds the notation $A+B$ for the union of A and B and AB, sometimes $D(A, B)$ for the intersection; they were not even called union and intersection but sum and product. When consulting the older literature, the reader should be aware of the notational differences. See for instance the classical book on set-theory by Hausdorff, *Grundzüge der Mengenlehre* (first edition 1914) [58] and the third revised edition from 1944 [59].[1] Even in 1952 Kuratowski used the notation $+$ and $\times$ [77]. In the first book of the Bourbaki series (*Théorie des ensembles, Fascicule de résultats*), published in 1939, one finds the notation $\cap$ and $\cup$ [11]. In logic Peano's notation was accepted by Whitehead and Russell in their *Principia Mathematica* (1913).

But there are more remarkable things in this introductory chapter, especially in the section on propositions. Peano observes that there are two possibilities: (i) propositions in which all elements are 'well determined' and (ii) propositions that contain 'variable elements'. If, in the last case, the proposition α contains the variables $x, y, z, \ldots$, Peano designs by $(x, y, \ldots): \alpha$ the set of all $(x, y, \ldots)$ for which α is true. If no confusion is possible about the variables, he agrees to omit the symbol $(x, y, \ldots):$ before the proposition.

Peano gave several examples for his notation and propositions. I quote some of them (l.c. p. 8, 10); see Appendix 1):

La scrittura $x: [f(x) = 0]$ rappresenta la classe di numeri x per cui $f(x)$ è nulla, cioè le radici dell'equazione $f(x) = 0$. (...)

La scrittura $x: [f(x) = 0] \cap x: [\varphi(x) = 0]$ rappresenta le radici comuni alle due equazioni $f(x) = 0$ e $\varphi(x) = 0$. (...)

$(x, y): [f(x, y) = 0]$ rappresenta l'insieme di tutte le coppie di valori di x ed y per cui $f(x, y) = 0$. (...)

$x: [f(x, y) = 0]$ rappresenta l'insieme dei valori di x per cui $f(x, y) = 0$; esso dipende dal valore di y (supposta data la funzione f). (...)

$$(x^2 - 3x + 2 > 0) = (x < 1) \cup (x > 2). \quad (\ldots)$$
$$[x: (x^2 + y^2 = 1) = \bigcirc] = (y < -1) \cup (y > +1).$$

---

1 In the third edition Hausdorff refers to Peano as to the notation $a \in A$ for 'a is an element of A'.

All this was written in 1888; but the mathematicians did not follow Peano in these notations. One may wonder why the mathematicians did not accept Peano's terminology. Or were they, in general, unfamiliar with it? This might be true because the book was seldom quoted. Bourbaki refers to it in the historical survey in his *Algèbre multilinéaire* (1948) [12], but then in view of the importance of this work of Peano with respect to the development of the theory of linear spaces and linear algebra. Earlier Church mentions this book by Peano in his extensive paper 'A bibliography of symbolic logic', published in 1936 [18], which contains a bibliography for the period 1666–1935.

The part on the theory of linear spaces and linear algebra is another remarkable aspect of Peano's book. He deals with it in chapter IX. The preceding chapters contain a rather naive geometrical calculus of lines, planes, volumes ('volumi'); some of the titles are 'Formazioni su d'una retta', 'Formazioni nel piano', 'Formazioni nello spazio'. He writes for instance (chapter I, p. 21): 'Siano *A*, *B*, *C*, ... dei punti nello spazio', but it is not clear what is to be understood by the word 'spazio' (space) because there is no definition of it. There is a definition of a line (line segment; linea *AB*) and Peano speaks about the 'grandezza *AB*' and, again, it is not clear what is meant by 'grandezza'. The concept of a vector is introduced (p. 37): 'Ogni formazione di prima specie della forma $B-A$ dicesi vettore.'[1] Bivectors are defined but it is all rather naive.

The most interesting part of the book is in chapter IX 'Trasformazioni di sistemi lineari', in which linear spaces are introduced nearly in the same way as we do nowadays. It is to be seen as a synthesis of the geometrical calculus of the preceding chapters.

I quote rather extensively some passages from this chapter so that the reader can make a comparison with modern definitions (see [96], p. 141; see Appendix 2):

72. Esistono dei sistemi di enti sui quali sono date le seguenti definizioni:

1. È definita l'*eguaglianza* di due enti a e b del sistema, cioè è definita una proposizione, indicata con a = b, la quale esprime

---

[1] 'Every formation of the first species of the form $B-A$ is called a vector.'

una condizione fra due enti del sistema, soddisfatta da certe coppie di enti, e non da altre, e la quale soddisfa alle equazioni logiche:

$$(a = b) = (b = a), \; (a = b) \cap (b = c) < (a = c).^1$$

2. È definita la *somma* di due enti a e b, vale a dire è definito un ente, indicato con a+b, che appartiene pure al sistema dato, e che soddisfa alle condizioni:

$$(a = b) < (a+c = b+c), \; a+b = b+a, \; a+(b+c) = (a+b)+c,$$

e il valor comune dei due membri dell'ultima eguaglianza si indicherà con a+b+c.

3. Essendo a un ente del sistema, ed *m* un numero intero e positivo, colla scrittura *m*a intenderemo la somma di *m* enti eguali ad a. È facile riconoscere, essendo a, b, ... enti del sistema, *m, n,* ... numeri interi e positivi, che

$$(a = b) < (ma = mb); \; m(a+b) = ma+mb;$$

$$(m+n)a = ma+na; \; m(na) = (mn)a; \; 1a = a.$$

Noi supporremo che sia attribuito un significato alla scrittura *m*a, qualunque sia il numero reale *m*, in guisa che siano ancora soddisfatte le equazioni precedenti. L'ente *m*a si dirà *prodotto* del numero (reale) *m* per l'ente a.

4. Infine supporremo che esista un ente del sistema, che diremo *ente nullo*, e che indicheremo con 0, tale che, qualunque sia l'ente a, il prodotto del numero 0 per l'ente a dia sempre l'ente 0, ossia

$$0a = 0.$$

Se alla scrittura a−b si attribuisce il significato a+(−1)b, si deduce:

$$a−a = 0, \; a+0 = a.$$

Def. *I sistemi di enti per cui sono date le definizioni 1, 2, 3, 4, in guisa da soddisfare alle condizioni imposte, diconsi* sistemi lineari.

Si deduce che se a, b, c, ... sono enti d'uno stesso sistema lineare, *m, n, p,* ... numeri reali, ogni funzione lineare omogenea della forma *m*a+*n*b+*p*c+ ... rappresenta un ente dello stesso sistema.

---

1 The symbol < means the implication.

Costituiscono sistemi lineari i numeri reali, e le formazioni della stessa specie nello spazio.

Costituiscono pure sistemi lineari le formazioni di prima specie su d'una retta, o nel piano, i vettori nel piano o nello spazio, e cosi via. Ma i punti dello spazio non costituiscono un sistema lineare, perchè le loro somme, secondo le definizioni date, non sono più punti, ma formazioni qualunque di prima specie.

73. Def. *Più enti* $a_1 a_2 \ldots a_n$ *d'un sistema lineare diconsi fra loro* dipendenti, *se si possono determinare n numeri* $m_1 m_2 \ldots m_n$, *non tutti nulli, in guisa che risulti*

$$m_1 a_1 + m_2 a_2 + \ldots + m_n a_n = 0.$$

In questo caso uno qualunque degli enti, il cui coefficiente non sia nullo, si può esprimere qual funzione lineare omogenea dei rimanenti.

Se gli enti $a_1 \ldots a_n$ sono fra loro indipendenti, e se fra essi passa una relazione

$$m_1 a_1 + \ldots + m_n a_n = 0, \text{ si deduce } m_1 = 0, \ldots, m_n = 0.$$

This leads Peano to the concept of dimension (l.c. p. 143):

Def. Numero delle dimensioni d'un sistema lineare è il massimo numero di enti fra loro indipendenti che si possono prendere nel sistema.[1]

There are some geometric examples, which he takes from the preceding chapters. Peano remarks that the dimension may be infinite. For finite-dimensional spaces he proves the existence of a basis (l.c. p. 143; see Appendix 3):

Teor. Se il sistema A è ad n dimensioni, presi nel sistema n enti indipendenti $a_1, \ldots, a_n$ e dato un nuovo ente a, si possono sempre determinare n numeri $x_1, \ldots, x_n$, in guisa che risulti

$$a = x_1 a_1 + \ldots + x_n a_n.$$

Inoltre essi sono determinati univocamente, ossia

$$(x_1 a_1 + \ldots + x_n a_n = x_1' a_1 + \ldots + x_n' a_n) = (x_1 = x_1') \cap \ldots \cap (x_n = x_n').$$

---

1 'Def. The number of the dimensions of a linear system is the maximal number of linearly independent objects in the system.'

THE DEVELOPMENT OF THE NOTION OF A LINEAR SPACE

The real numbers x_i are called the coordinates with respect to the system (a_i). There is even a notion of convergence in a finite-dimensional linear system: a sequence of elements is said to converge if all the sequences of the coordinates are convergent. For functions, defined on a linear system, this leads Peano to continuity and differentiability.

I quote the following examples of an infinite-dimensional space (l.c. p. 154; see Appendix 4):

> Si considerino le funzioni algebriche interi $f(x)$ d'una variabile numerica x. Intendendo con $f_1(x) = f_2(x)$ l'indentità dei valori di $f_1(x)$ e $f_2(x)$, qualunque sia il valori di x, con $f_1(x) + f_2(x)$ la funzione intera somma di $f_1(x)$ e $f_2(x)$, con $mf(x)$, ove m è un numero, il prodotto del numero m per la funzione $f(x)$, e con 0 una funzione nulla per ogni valore di x, le funzioni considerate sono enti di un sistema lineare. Se si considerano solo le funzioni di grado n, esse costituiscono un sistema lineare ad $n+1$ dimensioni; le funzioni intere di grado qualunque formano un sistema lineare ad infinite dimensioni.

Peano considered linear transformations; he defined them in a direct way, that is without introducing them by means of a matrix, as was customary still for years after his time. I give his definition (l.c. p. 145; see Appendix 5):

> Def. Un'operazione R, a eseguirsi su ogni ente a d'un sistema lineare A, dicesi distributiva, se li risultato dell' operazione R sull'ente a, che indicheremo con Ra, è pure un ente d'un sistema lineare, e sono verificate le indentità
>
> $$R(a+a') = Ra + Ra', \quad R(ma) = m(Ra),$$
>
> ove a e a' sono enti qualunque del sistema A, ed m un numero reale qualunque.

The sum and product of linear transformations are defined:

$$(R+S)a = Ra + Sa,$$

$$SR(a) = S(Ra).$$

The connection with matrices is established by means of the coordinates. The inverse R^{-1} is introduced.

This is a modern version of some parts of the theory of linear spaces, but unfortunately it was nearly forgotten. There is a review of Peano's *Calcolo geometrico* in the *Jahrbuch über die Fortschritte der Mathematik*, Bd. 20, p. 689–692 (1888) in which the reviewer mentions the general definition of linear systems, although he does not point it out as the most important element in the work; it seems that this review had also scarcely any influence on the further development.

In Peano's *Opere scelte*, Vol. II, the *Calcolo geometrico* was reprinted ([98] II, p. 3) but only as far as the introductory chapter on logic was concerned; the chapter on linear systems was omitted (see also Vol. III, p. 41).

In Felix Klein's *Vorlesungen über die Entwicklung der Mathematik im 19. Jahrhundert* II (1927) the Italian school is mentioned in the chapter 'Von der Entwicklung der Vektorlehre in den verschiedenen Ländern über Maxwells treatise hinaus' ([72], p. 48) but Klein's remark about Peano is not correct, and at least the merits of Peano on this point are not presented in the right way:

Hier sei nur bemerkt, dasz sich Peano in seinem Buche auf den Raum von 3 Dimensionen beschränkt und den Physikern so weit entgegenkommt dasz er die Bezeichnungen Vektor usw. aufnimmt.

We have seen on the contrary that Peano was familiar with the concept of an abstract infinite-dimensional linear space, even in the year 1888. Apparently Klein underestimated Peano.

In his book *A history of vector analysis* [19] Crowe mentions Peano and his *Calcolo geometrico* (p. 235, 236) as a mathematician promoting the ideas of Grassmann but he does not lay stress on the importance of chapter IX. Thus it remained unobserved that Peano gave a modern definition of the abstract concept of a linear space and that he was a predecessor of linear algebra. However, it is not the purpose of this book to give a history of linear algebra.

In the next chapter we will see that there is still another source for the general concept of a linear space: it is in Italy, where the Italian mathematicians, especially Pincherle, did important work on the introduction of linear spaces.

CHAPTER III

General analysis

§ 1 *The Italian school*

In the preceding pages I already mentioned the Italian mathematicians Pincherle and Volterra who studied functional operations and linear functionals at the end of the 19th century. In particular I considered in some detail Volterra and his 'fonctions de ligne', which led to the notion of a functional.

For an exposition of the course of the development it is important to comment further on their work. Of course I do not intend to give a complete survey of the work of the Italian mathematicians at the end of the 19th century; for instance I will not discuss Arzela and Dini and their importance for analysis. I will confine myself to Pincherle, who was a pioneer in functional analysis and the theory of functional operations.

Pincherle (1853–1936) started his publications in 1874; for a survey of his publications I refer the reader to the two volumes of *Opere scelte* [101]. Most of his publications are written in the Italian language.

His first publication dates from 1874; it was followed by several other publications and the 25th publication concerns functional operations 'Studi sopra alcune operazioni funzionali' ([101], t. I, p. 92) from which I quote (see Appendix 6):

Chiamo operazione funzionale qualunque operazione che eseguita sopra una funzione analitica dà per resultato una funzione analitica. Sono tali, per esempio, le operazione aritmetiche in numero finito, e per classi numerose di casi, anche in numero infinito, la derivazione e l'integrazione la risoluzione di equazioni finite o differenziali, la sostituzione ecc.

Fra gli algoritmi più notevoli per le operazioni funzionali va citata l'integrazione definita applicata ad una funzione di due variabili delle forma

$$\int_{(c)} f(x, y)\mathrm{d}y,$$

dove l'integrazione s'intende eseguita lungo una curva c, chiusa o no, del piano y.

As an example he mentions the formula of Cauchy

$$\varphi(x) = \frac{1}{2\pi i} \int_{(c)} \frac{\varphi(y)\mathrm{d}y}{y - x}.$$

This paper was followed by many other papers on functional operations; for instance 'Sur certaines opérations fonctionnelles représentées par des intégrales définies' ([101], t. I, p. 142) in which he considers integrals of the following type

$$\int A(x, y)\varphi(y)\mathrm{d}y$$

writing (p. 142): 'Je considère en effet l'expression ci-dessus comme un algoritme appliqué au *sujet* variable $\varphi(y)$ et dont les propriétés essentielles dépendent de la fonction $A(x, y)$.'

Furthermore 'Mémoire sur le calcul fonctionnel distributif' ([101], t. II, p. 1) in which he characterized 'le calcul fonctionnel' by saying 'on réunirait sous ce titre les chapitres de l'analyse où l'élément variable n'est plus le nombre, mais la fonction considérée en elle-même.' Distributive (linear) operators appeared earlier in 1897 in 'Appunti di calcolo funzionale distributivo' ([101], t. I, p. 388).

In a paper from 1896–1897 Pincherle deals with geometry in an infinite dimensional function space: 'Cenno sulla Geometria dello spazio funzionale' ([101], t. I, p. 368). Here he considers the set of all analytic functions in a variable x, writing (l.c. p. 368; see Appendix 7):

Ad un tale varietà, evidentemente ad un numero infinito di dimensioni, si può dare il nome di *spazio funzionale*; ogni serie di potenze di x sarà un *punto* di questo spazio ed i coefficienti della serie si potranno riguardare come le coordinate del punto.

He knows the concept of a linear space.

Essendo $c_1, c_2, ..., c_n$ numeri arbitrari ed $\alpha_1, \alpha_2, ..., \alpha_n$ funzioni linearmente indipendenti, l'insieme dei punti

$$c_1\alpha_1 + c_2\alpha_2 + ... + c_n\alpha_n$$

costituirà una varietà o spazio lineare ad $n-1$ dimensioni contenuto nello spazio funzionale.

In this paper Pincherle defines continuity of a distributive functional ([101], t. I, p. 370; see Appendix 8):

5.–Ricordiamo che una operazione A la quale applicata alle funzioni analitiche dà origine a funzioni pure analitiche, e che gode inoltre della proprietà distributiva, si dice *operazione funzionale distributiva*. Ognuna di queste operazioni dà pertanto una trasformazione dello spazio funzionale la quale, per ogni varietà lineare d'ordine finito di questo spazio, si riduce ad una omografia. Una tale operazione può essere *continua* per tutto lo spazio funzionale o per una parte (Γ) di esso. Dicendosi che A è continua per le funzioni di una classe (Γ) quando ad un numero g preso arbitrariamente corrisponde un numero h tale che alle funzioni della classe (Γ) minori in modulo di h per i valori della variabile compresi in un campo C, corrispondono funzioni che in un campo C' sono, in modulo, minori di g.

Essendo $\alpha(x, z)$ una funzione che per un campo continuo a due dimensioni, preso nel piano della variabile complessa z, appartiene alla classe (Γ), l'operazione A farà corrispondere alla curva $\alpha(x, z)$ una curva, ed in generale alle tangenti, ai piani osculatori, ... della prima corrisponderanno le tangenti, i piani osculatori, ... della seconda.

In a paper from 1906 'Sulle equazioni funzionali lineari' ([101], t. II, p. 264) Pincherle refers to the work of Hilbert, Schmidt and Hadamard. Of course he refers several times to the work of Volterra.

In these papers Pincherle studied *concrete* function spaces and *concrete* operators. But he must have known the definition of an abstract linear space, for in the paper 'Mémoire sur le calcul fonctionnel distributif' ([146], 1897) which I already mentioned, he refers to Peano's *Calcolo geometrico*, chapter IX and also to Grassmann. This is even more evident from a book Pincherle published in 1901 *Le operazioni distributive e le loro applicazioni all' analisi* [99].

In the introduction to this book he observes that for several years he has studied the problem of the inversion of integrals–which already was a subject of interest to Abel and Volterra–and the Laplace transformation, for which he refers to Poincaré.

Pincherle was particularly interested in the theory of functional operations for analytic functions. I quote from the introduction ([99], p. III; see Appendix 9):

> In primo luogo, osservando che ogni funzione analitica di una variabile è individuata dai valori attribuiti ad un numero generalmente infinito ma numerabile di parametri, si possono considerare quelle classi di funzioni che contengono tutte le combinazioni lineari dei loro elementi, ad esempio la totalità delle funzioni regolari nell'intorno di uno stesso punto, come spazî ad un numero generalmente infinito, ma numerabile di dimensioni.

He considers, for instance (p. 73), the linear space S^r of all analytic functions

$$\alpha(x) = a_0 + a_1 x + a_2 x^2 + \ldots$$

for which the radius of convergence is greater than a given positive number r, and the space S^∞ for which the radius of convergence is ∞. A large part of the book is occupied by applications to several parts of analysis, for instance differential equations and the Laplace transform. These chapters of a rather applied nature are preceded by four chapters in which Pincherle deals with the theory of linear spaces, introduced axiomatically. These chapters are entitled: I–L'insieme lineare generale ad n dimensioni, II–Generalità sulle operazioni, III–Radici e spazî di radici di una operazione distributiva, IV–Struttura degli spazi invarianti ad un numero finito di dimensioni.

His theory of linear space is entirely axiomatic; the starting point is a set of arbitrary, i.e. unspecified, elements. He refers to Laguerre and Peano and his axioms are almost identical to those of Peano. He introduced concepts as linearly independent or dependent vectors, dimension, coordinates and so on. Linear operators were introduced by him without reference to coordinates and he also defined the notion of the kernel of an operator ('radici e spazî di radici di una operazione distributiva'). Furthermore he studied invariant elements, that are elements α for which there is a number k, such that

$$A(\alpha) = k(\alpha)$$

(Eigenvectors, Eigenvalues). Pincherle remarks (p. 9) that it is useful to employ the geometric language.

This abstract theory is applied to analysis in the following chapters, for instance to the theory of differential equations.

For short, the chapters I to IV of this book, written in 1901, contain a systematic study of the theory of linear spaces. It seems that Peano's *Calcolo geometrico* and Pincherle's book were the first textbooks on this domain.

However, Pincherle's book too had only little influence: it was seldom quoted and I believe it was just ignored for many years after its publication. I will return to this fact in § 3.

I have been concerned here with the introduction of the concept of a linear space as an object of algebra. One should note, however, that the work of Volterra and Pincherle was particularly important for the development of functional analysis and general analysis.

§ 2 *General analysis*

In the section on integral equations I already mentioned that Volterra started the study of functionals in 1887, that is in his terminology 'fonctions de ligne' (p. 42 and following). Pincherle studied also functional operations. In this initial period both of them were occupied with *concrete* functional analysis, I mean they studied classes or spaces of real or complex valued functions and not an analysis with unspecified elements which is *abstract analysis*.

Although their work was fundamental for *general analysis*, *Fréchet* and *Moore* must be considered as the real pioneers in this field.

In 1904 Fréchet published a note in the *Comptes Rendus* [28] (see [33], p. 175) in which he discussed the property that a certain 'quantité dépendant de certains éléments (points, fonctions, etc. ...) atteint effectivement un minimum dans le champ considéré.' He observed that for certain properties in analysis it is not necessary to take into account the nature of the variables and that they can be established for elements of a set, satisfying certain adequate conditions, for example a topology defined axiomatically on the set so that it makes sense to speak of limits and of neighborhoods. This note was the starting point for the theory of abstract topological spaces and for general analysis.

In a lecture 'L'analyse générale et les espaces abstraits', delivered at the International Mathematical Congress held in Bologna in 1928 Fréchet gave the following description of general analysis (see [33], p. 5):

L'analyse fonctionnelle–et j'emploie ici l'heureuse locution introduite par M. Paul Lévy–a pour objet principal l'étude des propriétés infinitésimales des fonctions numériques dont la variable est, soit une ligne–et on a alors ce que M. Volterra appelle une fonction de ligne–soit une fonction ordinaire–et on a alors ce que M. Hadamard appelle une fonctionnelle.

Dans l'Analyse générale, la variable n'est plus nécessairement une ligne, ni une fonction ordinaire, ce n'est pas non plus une des variables de nature déterminée que la science amène à considérer comme l'argument d'une fonction. La variable n'est pas nécessairement une surface, ou une suite infinie de nombres, ou une transformation. C'est une variable abstraite. Mais il ne s'agit pas d'introduire ici quelque chose de mystérieux; si j'emploie cette expression, c'est simplement parce que c'est elle qui exprime le mieux l'idée que j'ai en vue.

Si, par exemple, je dis que je veux ajouter les nombres 8 et 3, vous ne m'accuserez pas d'ignorance parce que je n'ai pas spécifié que 8 est un nombre pair et 3 un nombre impair. Ce sont des détails que je connais, mais qui me sont indifférents au moment où je me dispose simplement à calculer la somme de ces nombres. Cette non-intervention de toutes les propriétés dans un problème déterminé est ce qui justifie la considération des éléments abstraits.

Un élément abstrait est, soit un élément dont la nature est indéterminée, soit un élément dont on connaît parfaitement la nature, mais dont, provisoirement, on n'a pas besoin de faire entrer la nature en ligne de compte.

La notion d'élément abstrait a cette utilité de ne faire intervenir la nature de la variable qu'au moment où cette intervention devient nécessaire et par suite de ne pas limiter à l'avance le champ d'application des résultats obtenus. Elle permet en outre d'éviter la répétition de raisonnements entièrement similaires dans une suite de théories parallèles, comme l'évite par exemple l'Analyse vectorielle pour les théories des forces, des vitesses, des moments, etc. ... Enfin, l'usage des éléments abstraits peut intéresser le philosophe, puisqu'il permet de mettre mieux en lumière le rôle de chacune des hypothèses qui entrent dans une démonstration.

Nous avons dit que l'Analyse générale étudie les fonctions numériques $y = f(x)$ d'une variable abstraite x. Mais une relation fonctionnelle $y = f(x)$ est aussi une transformation de la variable x dans la variable y. Pourquoi établir une distinction de principe entre x et y, pourquoi être général pour x et particulier pour y? Pourquoi y, au lieu d'être un nombre, ne serait-il pas lui aussi un être abstrait?

Et nous arrivons ainsi à la conception d'une science, l'Analyse générale, qui aurait pour objet l'étude des transformations $y = f(x)$ d'un être abstrait x en un être abstrait y.

In such a 'general analysis' it is necessary to define in an adequate way the notions of distance, neighborhood, limit and all what is necessary for building an analysis (compare footnote 1 on p. 45). 'Ordinary' analysis and the linear analysis I treated before are special cases of 'general analysis'. But the operations one is concerned with in general analysis are not necessarily linear. One studies, for instance, the concept of the derivative of a functional in connection with which I have to mention the name of the French mathematician Gateaux; compare the formula for the variation of a functional (Volterra) in chapter I, § 2.

In the lecture quoted above Fréchet gave the following motivation for the choice of his subject ([33], p. 5):

Si j'ai choisi pour sujet de cette conférence l'Analyse générale et la Théorie des ensembles abstraits, ce n'est pas seulement parce que c'est un des sujets qui ont le plus occupé mon attention. C'est surtout parce qu'il y a là une branche des sciences encore peu connue.

This was written in 1928, more than 20 years after the first steps were set in the domain of general analysis. The origin of general analysis even lies in the work of Volterra and Pincherle at the end of the 19th century. Fréchet remarked this in his lecture ([33], p. 5):

Et c'est bien ici le lieu de rappeler que l'Analyse générale n'aurait guère pu être même conçue sans les travaux des mathématiciens italiens et particulièrement de deux d'entre eux: MM. Pincherle et Volterra.

The growing importance of general analysis and functional analysis in these years is illustrated by the fact that two more lectures on this domain were delivered on this congress. Volterra lectured on 'La teoria dei funzionali applicata ai fenomeni ereditari'[1] [128] in which he dealt with 'fonctions de ligne' and functionals. The other lecture was given by Hadamard 'Le développement et le rôle scientifique du calcul fonctionnel' (reprinted in Hadamard's *Oeuvres*, t. 1, p. 435–453).

Speaking about the place of functional analysis in mathematics and the course of its development, Hadamard gave the following characterization (l.c. p. 437):

Il est clair qu'on sera conduit à opérer sur la fonction comme on a opéré sur le nombre, c'est à dire:

1) à regarder la fonction elle-même, non plus comme choisie une fois pour toutes, mais comme arbitrairement et continûment variable;

2) à lui faire subir les opérations les plus variées et les plus générales.

C'est la réalisation de ce programme qui s'appelle le *Calcul fonctionnel*.

Elle a commencé dès le principe même du Calcul infinitésimal. Les deux opérations fondamentales du Calcul infinitésimal, la différentiation et l'intégration, sont précisément des opérations fonctionnelles, à un degré très différent toutefois. La dérivée nous entraîne à peine en dehors du cercle des opérations algébriques, au point que, dans beaucoup de cas, tels que l'étude des racines multiples des équations, il soit à peine possible et assurément très illogique de séparer les deux ordres de problèmes. Il en va tout autrement pour la notion d'intégrale définie, qui fait intervenir toutes les valeurs de la fonction, à laquelle chaque élément de la courbe apporte sa part contributive; et là on peut même s'étonner que l'invention d'un pareil symbole n'ait pas conduit plus tôt les

---

[1] A phenomenon in mechanics or physics is called hereditary if the future of the system depends not only on the actual state of the parameters but also on all their values during a preceding interval. In the first case the problem leads to differential equations, in the second to integro-differential equations. For this terminology Volterra refers to Picard (see [127], p. 138).

géomètres à considérer les fonctions sous le point de vue de Dirichlet.[1]

Hadamard (*Oeuvres*, t. 1, p. 440) refers to Volterra and Pincherle, characterizing their work by the words:

> M. Pincherle s'est attaché particulièrement à la notion de *transmuée* (dont s'est occupé également notre regretté Bourlet, auquel nous empruntons d'ailleurs la dénomination qui précède), c'est à dire qu'il fait dépendre d'une fonction arbitraire prise comme argument une autre fonction résultat. M. Volterra, au contraire, considère ce que nous appelons aujourd'hui une *fonctionnelle*, c'est à dire déduit d'une fonction arbitrairement donnée un simple nombre: par exemple, la capacité électrique d'un conducteur, en tant que dépendant de la forme de ce conducteur.[2]

1 This reference to Dirichlet concerns the discussions on Dirichlet's definition of a function and the old definition of Euler (see [89]).

2 In a remarkable paper [13] Bourlet studied operations defined in such a way that to any function u, regular in a certain domain, corresponds another function of the same variable. He calls such an operation $\mathscr{E}$ a 'transmutation'. In the introduction he announces that his aim is to find the properties by which the operation of derivation is characterized. One of the results is that the derivation is characterized by the properties

$$\mathscr{E}(u+v) = \mathscr{E}u + \mathscr{E}v,$$

$$\mathscr{E}(uv) \quad = u\mathscr{E}v + v\mathscr{E}u,$$

joint to a condition of continuity.

Bourlet studies 'transmutations' satisfying general conditions such as

$$\mathscr{E}[\pi(u, v)] = \varphi(\mathscr{E}u, \mathscr{E}v),$$

for instance

$$\mathscr{E}(u+v) = \mathscr{E}u\mathscr{E}v.$$

The formula for the derivation of a product is reduced to this last formula by writing

$$\frac{\mathscr{E}(uv)}{uv} = \frac{\mathscr{E}u}{u} + \frac{\mathscr{E}v}{v},$$

and putting

$$Su = \frac{\mathscr{E}u}{u}.$$

There are several applications on the theory of differential equations.

As to Fréchet he remarked (l.c. p. 442):

Il était réservé à M. Fréchet–et, vers le même moment, à M. E. H.
Moore–de montrer que le plus simple et le plus clair était, cette
fois là, de savoir aller d'un coup jusqu'à l'extrème généralité et,
poussant jusqu'au bout l'abstraction mathématique, de raisonner
sur des éléments de nature entièrement indéterminée, les relations
entre ces éléments importants seules. (...) (for Moore see [90]).

Quant à la forme la plus générale à donner à l'expression d'une
variation, cette question se rattache en réalité à la conception de
vecteur dans les espaces fonctionnels ou abstraits, conception déjà
introduite dès les travaux de M. Pincherle, mais qui a dû être re-
prise, aux nouveaux points de vue auxquels se place aujourd'hui
l'Analyse fonctionnelle, par M. Fréchet et, d'une manière plus
approfondie, par MM. Norbert Wiener et Banach. (l.c. p. 440).

It would go too far to discuss the different kind of abstract topological
spaces that were introduced in general analysis; these are connected
with various definitions of the concepts of neighborhood and limit.
I refer the reader to the excellent survey by Fréchet [32], which contains an extensive bibliography.

It is worth observing that nowadays the term general analysis is no
longer used. In the modern terminology functional analysis comprises
the abstract general analysis as well as the concrete original functional
analysis.

In the next section I will make some general remarks on the way of
the development which I described in the preceding pages.

§ 3 *Critical remarks*

After the detailed exposition of the origin and the development of
functional analysis some more general remarks–and perhaps con-
clusions–should be made.

Looking back to the stream of papers that followed the first
trickling in the second half of the nineteenth century, one can distinguish two main branches. The relation between the two is rather
mysterious.

There is the first line, leading from Fredholm via Hilbert, Schmidt, Riesz, Helly to Hahn and Banach and the Polish school where the explosive development of functional analysis started. As we have seen these developments lie mainly in the period from 1900 to 1930, culminating in 1932, the year in which Banach's famous book was published.

There is a second approach from the Italian mathematicians Peano, Pincherle and Volterra, who refer in their work to Laguerre and Grassmann. Their work lies—at least as far as the introduction to functional analysis is concerned—mainly before 1900 and it has two culminating points: abstract linear spaces were already introduced in a formulation that does not essentially differ from the ones used at present by Peano in 1888 [96] and by Pincherle in 1901 [99].

Then follow in the first years of the 20th century the French mathematicians Hadamard and Fréchet whose work—as far as functionals and general analysis are concerned—is based on that of the Italian school.

Now the interesting point is the discussion of the relation between these lines of approach. In the line of Fredholm and Hilbert c.s. there is an evolution from concrete problems in classical analysis to the axiomatic theory of normed spaces, which was attained about 1920 (Banach). One observes this evolution in the theory of the systems of linear equations, in the integral equations and in the problem of moments.

The situation is about the same with Volterra and Pincherle; however, as we have seen before, the development took place earlier.

The curious fact is that there are practically no references to the Italian mathematicians in the work of Hilbert, Schmidt, Riesz and Helly; in his *Grundzüge* ([66], p. 2) Hilbert refers to a paper by Volterra, published in 1897. In his thesis [2] (1920) Banach refers to Pincherle among others. So, one guesses that Banach must have been acquainted with Pincherle's axiomatic definition of a linear space.

One cannot get rid of the idea that the results of the Italian school on the theory of linear spaces and linear operators have been lost and that the basic concepts would have to be rediscovered by the leading mathematicians about thirty years after the Italians had found them. It seems, indeed, that there was a vacuum in the theory of linear spaces in the years from 1900 (or 1888) to about 1920. Books from those years illustrate this fact.

The first textbook on *normed* linear spaces is Banach's *Théorie des opérations linéaires* (1932). But what about the books before 1930 that deal with the *algebraic* theory of linear spaces? These are the books which should be interesting in this respect.

Evidently the question is related to the history of the development of abstract algebra. An important concept in the theory of linear spaces, namely that of a set of linear independent elements, was, for instance, very well known, but it was not formulated in the language of linear spaces. Hankel, for instance, gives in his *Theorie der complexen Zahlensysteme* [56] (1867) a definition of this concept in the section in which he deals with formal systems composed of several 'imaginäre Einheiten'.

In Steinitz's paper 'Algebraische Theorie der Körper' (1910) [118] one reads (p. 183):

Ist L ein beliebiger Erweiterungskörper von K, so heiszen n Elemente $\alpha_1, ..., \alpha_n$ von L *linear* abhängig oder linear unabhängig in bezug auf K, je nach dem eine lineare homogene Relation

$$a_0\alpha_0 + ... + a_n\alpha_n = 0,$$

in welcher die Koeffizienten a_i Elemente aus K und nicht sämtlich gleich 0, sind, existiert oder nicht existiert.

It is not remarked that L is a linear space over K and that the concept of linearly independent elements is a notion which belongs to the domain of the theory of linear spaces. These are only examples; there are more indications in the work of the algebraists. For many years after Peano (1888) and Pincherle ($\pm$ 1890) defined linear transformations in an intrinsic way as linear (or distributive) operators, the mathematicians continued to introduce them in articles and books as linear substitutions, that is in the form of matrices. This is the case in older books, for instance in the well known textbook on algebra by H. Weber ([131], first edition 1896, II, p. 151; second edition 1899, II, p. 163), but also in more recent ones. The theory of abstract groups and fields was very well known in the early twenties, but it seems that this was not the case for the theory of linear spaces. Until the twenties –perhaps even later–one finds in books and articles the axioms of a linear space stated in all detail, as if this was a relatively new concept, although the axioms were known at the end of the 19th century. I mention for instance H. Weyl's *Raum, Zeit und Materie* [133]. There one finds (p. 15) all axioms of the concept of a linear space. Weyl refers to Grassmann, but not to Peano or Pincherle. Stone, in his book on the theory of Hilbert space [120], refers for his introduction of linear spaces to Weyl [133], and not to older sources. A substantial role in the development of the theory of linear spaces is played by Van der Waerden's books on 'modern' algebra [130]. In these books he refers to the courses of E. Artin and E. Noether.

I observe that in the period between 1900 and the twenties several books and articles on the theory of linear *algebra* and hypercomplex numbers were published; see for instance Dickson's book on linear algebra, edited in 1914 [21]. But they don't mention the geometrical aspects. See also Dickson *Algebras and their arithmetics* [22], chapter II 'Linear sets of elements of an algebra'. I mentioned only some books by way of example and I did not systematically investigate the literature on the development of the theory of linear spaces. The reader will find an excellent bibliography in Wedderburn's book *Lectures on matrices* [132], beginning in 1853 with Hamilton's *Lectures on quaternions* [55] and systematically proceeding until 1933. See also a paper by Hawkins [60].

Crowe writes in the preface of his book *A history of vector analysis* that he was hindered by the absence of studies of the history of related areas such as complex numbers and linear algebra ([19], p. VIII).

A history of the development of the theory of linear spaces is certainly desirable; in particular it would be most interesting to know more about the place of E. Artin and E. Noether in the development.

A main feature of the development I described in the preceding pages was the strong influence of algebra on analysis: it seems that the cooperation of analytical, algebraic and topological methods marked the great progress. It is not only in analysis that an algebraization may be observed. The algebraization of geometry is also a feature of modern mathematics. I have only to mention modern algebraic geometry. The history of algebraization of geometry into modern times has, as far as I know, not yet been written.

CHAPTER IV

Some characteristic topics

In this chapter I will give some examples of the tendency towards algebraization of modern analysis and of the application of functional-analytic methods on problems of classical analysis. The presentation is mainly expository; I must refer the reader to other books and to the original papers.

§ 1 *The theorem of Weierstrass*

In the classical form the theorem states that any real continuous function defined on a closed interval $[a, b]$ can be approximated uniformly by means of polynomials with rational coefficients or, equivalently, any such function can be written in the form of a uniformly convergent series of polynomials with rational coefficients. This is more or less the exact formulation of the vague Eulerian definition of a continuous function as an analytical expression (for the history of the development of the concept of function see my paper [89]). In the course of time several proofs of this classical theorem were given; it was not considered a simple theorem.

This theorem has been generalized, on the one hand by weakening the condition of considering functions defined on an interval, on the other hand by allowing approximation by other functions than polynomials. One even considers not only real-valued functions, but functions taking their values in a valued field K.

For the generalized theorem the notion of an *algebra* (see chapter I, § 5) is fundamental.

Let S be a compact topological space and $C(S)$ the linear space of the real-valued continuous functions defined on S. $C(S)$ is a commutative normed algebra if the norm of $f \in C(S)$ is defined by

$$\|f\| = \sup_{x \in S} |f(x)|.$$

Then one has the following generalization of the theorem of Weierstrass:

Theorem of Stone-Weierstrass. Let A be a subalgebra of $C(S)$ that contains the constant functions and separates the points of S.[1] Then A is dense in $C(S)$.

Taking for S the interval $[a, b]$ and for A the set of all polynomials, one obtains the classical theorem. In this general form it is a rather simple theorem and even the proof is simple. Moreover, in this form one realizes that only rather trivial properties of the polynomials are used.

The development of this theorem in the course of time gives an illustration of the three phases, which Volterra indicated in the lecture which I mentioned in the introduction. In the general form the theorem has reached the third phase, the 'cadre didactique'.

§ 2 *Banach algebras*

The introduction of algebras in analysis entails the use of algebraic methods, for instance the theory of ideals. One is interested in the relations between the analytical and the algebraic structure; this is one of the characteristic features of modern analysis in contrast with classical analysis.

In classical analysis one studies, for instance, the individual properties of the entire functions – the distribution of the values, the behaviour when $z \to \infty$ and so on.

In functional analysis one is interested in the properties of the family of entire functions as a linear space over the complex numbers, for instance the dual space and relations with other function spaces.

Let us, for example, consider the Banach algebras; a Banach algebra is a normed algebra which is complete for the norm topology.

Let A be a commutative Banach algebra over the complex number

---

[1] A separates the points of S if for each $x, y \in S$, $x \neq y$, there is $f \in A$ such that $f(x) \neq f(y)$. Evidently the polynomials satisfy this condition.

field; suppose there is a unit element e in A with $\|e\| = 1$. One of the problems in studying these algebras is to give a characterization. That is to say: can one give a survey of all these algebras? Under some restrictions this is possible indeed. Under certain supplementary conditions any such algebra can be identified (up to isomorphisms) with an algebra of complex valued functions defined on an adequate space. The following rough sketch may serve to give the reader an impression.

One considers the collection Δ of the maximal ideals in A. Suppose the intersection of all the maximal ideals of A consists of only one element: the 0-element. The algebra is then called semisimple. Now, suppose that A is semisimple. There is a topology on Δ, making Δ a compact topological space. The Banach algebra $C(\Delta)$ of the complex valued continuous functions on Δ is introduced and one proves that there is a isomorphism from A into $C(\Delta)$

$$\Phi: x \to \varphi_x, \quad x \in A, \quad \varphi x_x \in C(\Delta).$$

This leads to the theorem:

Any commutative semisimple Banach algebra with unit is algebraically isomorphic with an algebra of continuous functions on Δ. This is the so called *Gelfand representation*.

If A is a C^*-algebra–that is an algebra on which an involution is defined (an involution is a mapping from A into A with similar properties as the complex conjugation in the complex number field)– one even proves that the algebra is isometrically isomorphic with the algebra of *all* continuous functions on Δ.

This theorem does not mean that any such Banach algebra is identical with a function algebra in the common sense. There is an isomorphism between the algebra and the corresponding function algebra and thus, roughly said, both have the same properties and cannot be distinguished from each other if one considers only these properties.

The corresponding result for non-commutative C^*-algebras is less simple. Every such algebra is isometrically isomorphic with a sub-algebra of the algebra of all bounded linear operators on some complex Hilbert space.

Example. Consider the set $C(I)$ of the real continuous functions defined on the interval $I = [a, b]$. Evidently $C(I)$ is a ring. It is trivial that for any fixed $x_0 \in I$ the set

$$\{f \in C(I) | f(x_0) = 0\}$$

is a maximal ideal in $C(I)$.

The question arises whether, conversely, every maximal ideal in $C(I)$ can be obtained in this way. The answer is affirmative. This is proved by means of the continuity of the functions and the compactness of I. For the extensive theory of Banach algebras see [92].

§ 3 *Haar measure*

It is trivial that the Lebesgue measure on $\mathbb{R}$ is invariant under translations. This property expresses a connection between Lebesgue measure and the structure of $\mathbb{R}$ as an additive group. It has considerably been generalized.

Let G denote a locally compact topological group and let $C(G)$ be the space of the real-valued continuous functions on G with compact support. For any $y \in G$ and $f \in C(G)$, the function $fy \in C(G)$ is defined by

$$(fy)(x) = f(yx), \quad x \in G.$$

Let μ be a measure (integral) defined for any $f \in C(G)$; the value of the integral for any f is denoted by $\mu(f)$ or $\int f d\mu$. For any $y \in G$, define a measure $y\mu$ by

$$(y\mu)(f) = \mu(fy), \quad f \in C(X).$$

If a measure $\mu \neq 0$ satisfies the relation $y\mu = \mu$ for every $y \in G$, it is called a left invariant Haar measure.

One proves that there exists such a measure on any locally compact group G. This property was the starting point of an analysis on topological groups in which algebraic and analytic features are connected with each other; the ordinary real analysis is a special case of this analysis (see [85]).

For an outline of the development of the integral see [20].

§ 4 *Germs of functions*

Algebraic methods are used in the study of local properties of functions. As an example I take the theory of the functions of several complex variables (which is considerably more difficult than the theory of functions of one complex variable).

Let $\mathbb{C}$ be the field of complex numbers. If

$$z = (z_1, ..., z_n) \in \mathbb{C}^n = \mathbb{C} \times \mathbb{C} \times ... \times \mathbb{C},$$

one defines

$$|z| = \max(|z_j|; \ j = 1, 2, ..., n)$$

The neighborhoods of z are designed by $U, V, W, ...$

Consider functions $f, g, ...: U \to \mathbb{C}$. By f_U, g_V one means that f, resp. g are defined on U, resp. V. If $V \subset U$, the restriction of f_U to V is denoted by $f_U|V$.

Definition. f_U, g_V are called equivalent in z if there is an open neighborhood W of z, $W \subset U \cap V$, such that

$$f_U|W = g_V|W.$$

It is easily proved that this is an equivalence relation. Any function f, defined in an open neighborhood of z, belongs to an equivalence class; this class is called the *germ* of f in z. Thus, the germ of a function in z depends not only on the value of the function in z, but also on the values in a neighborhood of z. On the set of all germs in a point z one defines in an evident way the structure of a ring.

Putting conditions on the functions, for instance continuous functions or holomorphic functions, one obtains the ring of the germs of continuous resp. holomorphic functions in z. The ring of the germs of holomorphic functions in z is denoted by $_n O_z$. This ring is isomorphic with the ring of the convergent power series in z. For $z = (0, 0, ..., 0)$ the ring is denoted by $_n O$. The problem is to study the properties of this ring. This is done by algebraic methods. For example:

(i) $_n O$ *is an integral domain* (this is a consequence of the identity theorem for holomorphic functions). The quotient field is the field of meromorphic functions.

(ii) *The ring $_n O$ is noetherian, i.e. every ideal in $_n O$ is finitely generated.* For this theory see [50].

§ 5 *Non-archimedean analysis*

The theory of Banach spaces and Banach algebras over fields different from $\mathbb{R}$ and $\mathbb{C}$ (namely non-archimedean valued fields, i.e. valued fields for which the valuation satisfies the strong inequality $|x+y| \leq$ $\leq \max(|x|, |y|)$, instead of the usual triangle-inequality) can be treated, as well as the theory of holomorphic functions in such a field. As to the latter I mention that the classical theory of analytic continuation does not lead to any result if one tries to follow the classical procedure of complex analysis. The main reason lies in the circumstance I mentioned in footnote 1 on p. 73. It is remarkable that nevertheless one has succeeded in giving a satisfactory theory of analytic functions for this case with the aid of algebraic methods which thus seem to be more powerful in the domain of analysis than topological and analytical methods alone. It is the cooperation of algebraic and analytical methods which yields results in these cases. For these theories see the bibliography in [88].

§ 6 *Classical problems*

From the beginning of the development of functional analysis Banach, Saks and Steinhaus have applied the methods of functional analysis to classical problems. I mention problems of the existence of functions with various singularities.

(i) *Continuous functions without derivative.* With the methods of functional analysis Banach proved the existence of continuous functions which don't have a derivative on a set of positive measure.

This is proved in the following way. Let C be the linear space of the continuous functions defined on [0, 1]. It is a Banach space in the usual way. C is supposed to be embedded in the metric space S of the measurable functions. For any $f \in C$ and $h \neq 0$ one defines the function φ_h by

$$\varphi_h(t) = \frac{f(t+k)-f(t)}{h}, \quad 0 \le t \le 1.$$

Thus a linear mapping T_h from C into S is defined by putting

$$T_h(f) = \varphi_h.$$

Now, if f' exists almost everywhere for all $f \in C$, $\lim_{h \to 0} T_h(f)$ existed for all f, where the limit is to be considered as a limit in S, that is to say a *limit in measure*. Putting

$$T(f) = \lim T_h(f),$$

T is a linear operator from C into S, and it follows from the theory of operators that T is continuous. This implies that if (f_n) is a sequence of continuous functions, uniformly tending to 0, $T(f_n)$ tends in S to 0. Now, this is contradicted by the example

$$f_n : f_n(t) = \frac{1}{n} \sin \frac{nt}{2\pi},$$

since (f_n') does not tend in measure to 0. For some more examples see Steinhaus [117].

(ii) *Continuous functions whose Fourier series diverges in the points of certain sets.* It was a classical problem in analysis whether such functions exist. Banach treated this problem with the methods of functional analysis (see [3]). He studied the more general case of the development of a function with respect to an orthogonal sequence of quadratic integrable functions (φ_n). These are series of the form

$$f(t) = \sum_{k=1}^{\infty} a_k \varphi_k(t),$$

where the a_k, the Fourier coefficients, are defined by

$$a_k = \int_0^1 \varphi_k(t) f(t) dt.$$

It is proved, for instance, that there are continuous functions for which the Fourier series is divergent in a certain set of points $(t_n)_{n \in N}$. This is proved again by means of functional operators

$$T_n(f) = \sum_{k=1}^{n} a_k \varphi_k;$$

the convergence of the sequence of these operators is studied. To get the result a method is used, due to Hankel, by which it is possible to construct, starting from objects having only one singularity (for instance a continuous function whose Fourier series diverges in one point), an object which has an infinite number of singularities; see Hankel [57], in particular the annotations in *Ostwald's Klassiker* with reference to a paper by Cantor on this method; see also Banach-Steinhaus [5]. Hankel called this method 'Methode der Kondensation der Singularitäten'. This method has been considerably generalized. The following theorem is proved:

Let be given a Banach space X and a sequence of normed linear spaces Y_n $(n = 1, 2, ...)$. Let (T_{nm}) $(m = 1, 2, ...)$ be a sequence of bounded linear operators from X into Y_n. Suppose for each n there exists $x_n \in X$ such that

$$\overline{\lim_{m \to \infty}} \| T_{nm} x_n \| = \infty.$$

Then the set

$$\{x \in X | \overline{\lim_{m \to \infty}} \| T_{nm} x \| = \infty \text{ for all } n = 1, 2, ...\}$$

is of the second category.[1]

This is nearly the form in which it was given by Banach (see [4], p. 24; [138], p. 74; Saks [111]). For Hankel's work see [145].

These methods don't provide the means of defining particular functions with the desired singularities, as is done in classical analysis by the construction of examples. On the other hand they permit to describe the properties of the spaces of all the functions with these singularities, for instance one gets theorems like: 'most' continuous functions are nowhere differentiable. For further information see [4], [138].

---

1 A set A in a space E is called of the first category if it is the countable union of sets which are nowhere dense in E. Sets which are not of the first category are called of the second category.

§ 7 *Some fundamental theorems in functional analysis*

In chapter I I treated the history of the theorem of Hahn-Banach, which marked the beginning of the explosive development of modern functional analysis. At the end of that chapter I mentioned some other fundamental theorems. I give here these theorems; for the proof I must refer the reader to the textbooks on functional analysis (see for instance [138]).

(i) *The uniform boundedness theorem* (*Banach-Steinhaus*). Let X and Y be Banach spaces. Let T_a, $a \in A$, be a family of continuous linear operators from X into Y such that

$$\sup_{a \in A} \| T_a x \| < \infty$$

for all x $\in X$. Then

$$\sup_{a \in A} \| T_a \| < \infty.$$

This theorem allows to conclude the boundedness of the family $(T_a)_{a \in A}$ from the boundedness of $(T_a x)_{a \in A}$ for every x separately. It is an important theorem for applications of the theory; it can be used for proving the classical problem (ii) mentioned in § 6. The theorem was already proved in the beginning of the theory (see [5]).

(ii) *The open mapping theorem* (*Banach*). Let X and Y be Banach spaces. Let T be a continuous linear operator from X *onto* Y. Then T maps every open set of X onto an open set of Y, this means that T is a homeomorphism.

This theorem too dates from the beginning of the theory (see [4]). The form in which I give it here is not the most general form; the conditions on X and Y may be weakened. Note that in general a continuous function does not map open sets onto open sets, but under the condition of this theorem it does. The proof of the theorem requires topological tools (the so called category proofs).

An easy application of this theorem leads to the third fundamental theorem.

(iii) *The closed graph theorem* (*Banach*). Let X and Y be Banach spaces. Let T be a linear operator from X into Y. The *graph* of T is defined as the subset

$\{(x, Tx)|x \in X\}$

of the product space $X \times Y$. In a natural way a norm can be defined on $X \times Y$, making this product space to a Banach space. T is called a *closed linear operator* when its graph is a closed linear subspace of $X \times Y$.

Theorem. A closed linear operator T from X into Y is continuous. The conditions on X and Y may also be weakened (see [4]).

Final remarks

At the end of this book there is a place for some philosophy. In the introduction I mentioned a tendency towards algebraization of analysis and from the preceding pages the reader may have got the impression that algebraization is the highest wisdom in mathematics. Therefore some more clearness on this point is desirable.

The term 'algebraization' may be somewhat misleading in itself. There was a time that the line between algebra and analysis was drawn at the introduction of limits. Most of the classical text of Van der Waerden's *Moderne Algebra* [130] satisfies this criterion of belonging to algebra. However limitprocesses have penetrated into algebra and in some branches, e.g. topological groups, Lie groups, a succesful amalgamation of algebra and analysis (or alternatively topology) is the very essence. An attempt from the side of logic to characterize algebra as the study of certain first order theories is in the present context not illuminating at all.

I also mentioned in the introduction a shift in attention from the original objects (such as numbers, vectors, equations) to structures (such as groups, fields, vector spaces). This shift seems to be the main characteristic of the modern approach to mathematics. The structures are made the object of study, one looks for closure properties, homomorphisms, substructures, etc. The present monograph is a history of the triumphs of this latter technique of 'algebraization'. However hand in hand with the replacement of the object by the structure (most strikingly revealed in category theory) goes a demand for strong abstract means of proof. The very abstraction from the original object annihilates the intrinsic means of proof inherent to the object under consideration. As a consequence various strong principles have to be adopted, for instance Zorn's lemma which is highly inconstructive. In itself there is no reason to reject this, as even in everyday analysis some use of the axiom of choice is made. It is rather the fact that smooth and abstract methods are bought at the price of stronger and stronger axiom systems.

Bishop, in his book *The foundations of constructive analysis*, has shown how to reconcile the methods of modern functional analysis and the demands of a concrete and effective mathematics. See also a monograph of Stolzenberg [143].

Much earlier Brouwer had undertaken to reconstruct parts of mathematics, including subjects as measure theory, thereby isolating the constructively significant parts of mathematics.

Roughly speaking one can say that in the objects of functional analysis all specific details of the original objects are obscured. An instructive example is provided by the theorem of Stone-Weierstrass.

The possibility of approximating continuous functions by means of polynomials has become a simple problem by means of the general theorem of Stone-Weierstrass. But the proof does not provide the means of constructing approximating polynomials. Several classical proofs – many were given in the course of time – are constructive in the sense that they not only prove the existence but also permit to define the approximating polynomials.

Another example is in potential theory. Classical potential theory can roughly be described as the theory of the partial differential equation $\Delta u = 0$. In axiomatic potential theory one studies certain linear spaces of functions, defined on a topological space, and in an abstract way called 'harmonic functions'. This theory (which contains the classical theory as a special case) has nothing to do with differentiation. For many theorems from the classical theory differentiability is shown to be superfluous, and thus a theory with a larger scope is obtained. However axiomatic potential theory is of no use if one wants to have explicit solutions of the equation of Laplace, which sometimes may be necessary.

The advance of numerical analysis and computational mathematics has again opened the eyes of mathematicians to the problems of constructive mathematics, including functional analysis. In conclusion one can say that functional analysis is a paradigma of the twentieth century trend towards abstraction and axiomatization, but that at the same time the classical constructive traditions have kept their value.

Appendix

1. Ad p. 116

'The notation $x:[f(x) = 0]$ represents the class of the numbers x for which $f(x)$ is zero, i.e. the roots of the equation $f(x) = 0$. (...)

The notation $x:[f(x) = 0] \cap x:[\varphi(x) = 0]$ represents the common roots of the two equations $f(x) = 0$ and $\varphi(x) = 0$. (...)

$(x, y): [f(x, y) = 0]$ represents the set of all the couples of values of x and y for which $f(x, y) = 0$. (...)

$x: [f(x, y) = 0]$ represents the set of the values of x for which $f(x, y) = 0$; this set depends on the value of y (the function f is supposed to be given).'

2. Ad p. 117

'72. There exist systems of objects on which the following definitions are given:

1. The *equality* of two objects of the system is defined, i.e. a proposition is defined, denoted by $a = b$, expressing a condition for two objects of the system, which is satisfied by certain couples of objects, and not by other, such that the following logical equations are valid:

$(a=b) = (b=a)$, $(a=b) \cap (b=c) < (a=c)$.

2. The *sum* of two objects a and b is defined, i.e. an object is defined, denoted by $a+b$, also belonging to the system, which satisfies the conditions:

$(a=b) < (a+c = b+c)$, $a+b = b+a$, $a+(b+c) = (a+b)+c$,

and the common value of the last equality is denoted by $a+b+c$.

3. If a is an object of the system and m a positive integer, then we understand by ma the sum of m objects equal to a. It is easy to see that for objects $a, b, \ldots$ of the system and positive integers $m, n, \ldots$

one has

$$(a=b) < (ma=mb); \quad m(a+b) = ma+mb;$$

$$(m+n)a = ma+na; \quad m(na) = mna; \quad 1a = a.$$

We will suppose that for any real number m the notation ma has a meaning such that the preceding equations are valid. The object ma will be called the *product* of the (real) number m and the object a.

4. Finally we will suppose that there is an object of the system, called the *zero-object* and denoted by 0, such that for any object a the product of the number 0 and the object a always equals the zero-object, i.e.

$$0 \cdot a = 0.$$

If by the notation $a-b$ one means $a+(-1)b$, then one shows

$$a-a = 0, \quad a+0 = a.$$

Def. The systems of objects for which the definitions 1, 2, 3, 4 are given such that the preceding conditions are satisfied, are called linear systems.

One deduces that, if a, b, c, ... are objects of the same linear system and m, n, p, ... are real numbers, every homogeneous linear function of the form $ma+nb+pc+...$ is an object of the same system.

Linear systems are the real numbers and the formations of the same kind in space.

Linear systems are also formations of the first kind on a line, or in the plane, the vectors in the plane or in space etc. The points of the space, however, do not form a linear system, because their sums, following the definitions, are no longer points but arbitrary formations of the first kind.

73. *Def. Several objects $a_1, a_2, ..., a_n$ of a linear system are called independent, if n numbers $m_1, m_2, ..., m_n$ not all zero, can be determined such that*

$$m_1 a_1 + m_2 a_2 + ... + m_n a_n = 0.$$

In this case every arbitrary object, whose coefficient is not zero, can be written as a homogeneous linear function of the other objects.

If the objects $a_1, ..., a_n$ are independent, and if there is a relation $m_1 a_1 + ... + m_n a_n = 0$, then one has $m_1 = 0, ..., m_n = 0$.'

3. Ad p. 119

'Theorem. If the system A has n dimensions and one takes n independent objects $a_1, ..., a_n$ in the system, then for any given new object a one can determine n numbers $x_1, ..., x_n$ such that

$$a = x_1 a_1 + ... + x_n a_n.$$

Moreover these numbers are uniquely determined, i.e.

$$(x_1 a_1 + ... + x_n a_n = x'_1 a_1 + ... + x'_n a_n) = (x_1 = x'_1) \cap ... \cap (x_n = x'_n).'$$

4. Ad p. 120

'Let us consider the entire algebraic functions $f(x)$ of a numerical variable x. If we design by $f_1(x) = f_2(x)$ the equality of the values of $f_1(x)$ and $f_2(x)$ for any arbitrary value of x, by $f_1(x) + f_2(x)$ the entire function which is the sum of $f_1(x)$ and $f_2(x)$, by $mf(x)$, where m is a number, the product of the number m and the function $f(x)$, and by 0 a function which is zero for every value of x, then these functions are the objects of a linear system. If one considers only the functions of degree n, then these functions form a linear system with $n+1$ dimensions; the entire functions of arbitrary degree form a linear system with infinitely many dimensions.'

5. Ad p. 120

'Def. An operation R, to carry out on every object a of a linear system A, is called distributive if the result of the operation on the object a, which we shall design by Ra, is again an object of the linear system and if the identities

$$R(a + a') = Ra + Ra', \quad R(ma) = mR(a),$$

in which a and a' are arbitrary objects of the system A and m an arbitrary real number, are satisfied.'

6. Ad p. 123

'I call a functional operation every operation, which, when executed on an analytic function, has an analytic function as result. Examples are the finite and, in many cases, also the infinite arithmetic operations, the differentiation and the integration, the solution of finite equations or differential equations, the substitution, etc.

Among the most important algorithms for functional operations must be mentioned the definite integration, applied to a function of two variables, of the form

$$\int_{(C)} f(x, y) \, dy,$$

in which the integration is supposed to be taken along a curve C, closed or not, in the y-plane.'

7. Ad p. 124
'Such a variety, which has evidently infinitely many dimensions, can be given the name *function space*; then every power series in x is a *point* of this space, and the coefficients of the series can be considered as the coordinates of the point. If c_1, c_2, ..., c_n are arbitrary numbers and α_1, α_2, ..., α_n linearly independent functions, then the set of the points

$$c_1\alpha_1 + c_2\alpha_2 + \ldots + c_n\alpha_n$$

forms a variety or linear space of $n-1$ dimensions which is contained in the function space.'

8. Ad p. 125
'We recall that an operator A such that the application on an analytic function generates again an analytic function and which has the distributive property, is called a distributive functional operation. Each of these operations generates a transformation in the function space, which induces a homography on every linear variety of finite order. Such an operation may be *continuous* on the entire space or on a part (Γ) of it. We say that A is continuous for the functions of a class (Γ) if to any arbitrary number g there corresponds a number h such that to the functions of the class (Γ) which have a modulus smaller than h for the values of the variable, belonging to a domain C, correspond functions which have a module smaller than g in a domain C'. Is $\alpha(x, z)$ a function, belonging to the class (Γ) for a continuous domain of two dimensions in the plane of the complex variable z,

then the operation A associates to the curve $\alpha(x, z)$ a curve, and in general the tangents, the osculation planes, ... of the first correspond to the tangents, the osculation planes, ... of the second.'

9. Ad p. 126

'Because any analytic function of one variable is determined by the values attributed to in general infinitely but countably many parameters, we can consider in the first place as spaces with in general infinitely but countably many dimensions the function classes which contain all linear combinations of their elements, for example the total of the functions which are regular in the neighborhood of the same point.'

Bibliography

[1] Abel, N. H., Oplösning af et Par opgaver ved Hjelp af bestemte Integraler, *Magazin for Naturvidenskaberne*, Aargang 1823, 3 Hefte, 55–68 (= Solution de quelques problèmes à l'aide d'intégrales définies, *Oeuvres complètes*, t. 1, pp. 11–27, Christiania 1881.

[2] Banach, St., Sur les opérations dans les ensembles abstraits et leur application aux équations intégrales, *Fund. Math.* t. III, 123–181.

[3] Banach, S., Sur les fonctionnelles linéaires II, *Studia Mathematica* I, 223–239 (1929).

[4] Banach, S., *Théorie des opérations linéaires*, Warzawa 1932.

[5] Banach, S. and H. Steinhaus, Sur le principe de la condensation de singularités, *Fund. Math.* 9, 50–61 (1927).

[6] Bernkopf, M., The Development of Function Spaces with Particular Reference to their Origin in Integral Equation Theory, *Archive for History of Exact Sciences* 3, 1–96 (1966/1967).

[7] Bernkopf, M., History of Infinite Matrices. A Study of Denumerably Infinite Linear Systems as the First Step in the History of Operators Defined on Function Spaces, *Archive for History of Exact Sciences* 4, 308–358 (1967/1968).

[8] Bois-Reymond, P. du, Bemerkungen über $\Delta z = \dfrac{\partial^2 u}{\partial x^2} + \dfrac{\partial^2 u}{\partial y^2} = 0$, *Journal f. reine und angew. Math.* 102, 204–229 (1888).

[9] Bolzano, B., *Oeuvres*, t. 5, Prague 1948.

[10] Bottema, O., Het eeuwfeest van een ongelezen boek, *De Gids* 109, 160–173 (1946).

[11] Bourbaki, N., *Théorie des ensembles, Fascicule de résultats*, Paris 1939.

[12] Bourbaki, N., *Algèbre multilinéaire*, Paris 1948.

[13] Bourlet, M. C., Sur les opérations en général et les équations différentielles linéaires d'ordre infini, *Ann. Ec. Norm. Sup.* (3) 14, 133–190 (1897).

BIBLIOGRAPHY

[14] Boyer, C. B., *A History of Mathematics*, New York etc. 1968.

[15] Carvallo, E., Sur les systèmes linéaires, les calculs symboliques différentiels et leur application à la physique mathématique, *Monatsh. für Math. und Physik* II, 177–216, 225–266, 311–330 (1891).

[16] Cauchy, A., Sur les clefs algébriques, *Oeuvres*, t. XI, 1e série, 439.

[17] Cauchy, A., Sur les avantages que présente dans un grand nombre de questions l'emploi des clefs algébriques, *Oeuvres*, t. XII, 1e série, 21.

[18] Church, A., A bibliography of symbolic logic, *The Journal of Symbolic Logic*, Vol. 1, 121–218 (1936).

[19] Crowe, M. J., *A history of vector analysis*, Notre Dame, London, 1967.

[20] Dalen, D. van and A. F. Monna, *Sets and Integration, an outline of the development*, Groningen 1972.

[21] Dickson, L. E., *Linear algebras*, Cambridge 1914.

[22] Dickson, L. E., *Algebras and their arithmetics*, New York 1938.

[23] Dieudonné, J., La dualité dans les espaces vectoriels topologiques, *Ann. Ec. Norm. Sup.* (3) 59, 107–139 (1942).

[24] Dinghas, A., Ehrhard Schmidt (Erinnerungen und Werk), *Jahresber. Deutschen Math. Verein.* 72, 3–17 (1970).

[25] Doetsch, G., *Einführung in Theorie und Anwendung der Laplace-Transformation*, Basel 1958.

[26] Eggleston, H. G., *Convexity*, Cambridge 1963.

[27] Fourier, J., *Théorie analytique de la chaleur*, Paris 1822.

[28] Fréchet, M., Généralisation d'un théorème de Weierstrass, *C. R. Ac. Sci. Paris*, t. 139, 848–849 (1904).

[29] Fréchet, M., Sur quelques points du calcul fonctionnel, Thèse Paris 1906; *Rend. Circ. Mat. Palermo*, t. 22, 1–74 (1906).

[30] Fréchet, M., Sur les ensembles de fonctions et les opérations linéaires, *C. R. Ac. Sci. Paris*, t. 144, 1414–1416 (1907).

[31] Fréchet, M., Sur la notion de voisinage dans les ensembles abstraits, *Bull. Sc. Math.*, t. XLII, 1–19 (1918).

[32] Fréchet, M., *Les espaces abstraits*, Paris 1928.

[33] Fréchet, M., *Pages choisies d'analyse générale*, Paris 1953.

[34] Fredholm, I., Sur une nouvelle méthode pour la résolution du problème de Dirichlet, *Ofversigt af Kongl. Vetenskaps-Akademiens Förhandlingar*, 1900, no. 1, Stockholm.

[35] Fredholm, I., Sur une classe d'équations fonctionnelles, *C. R. Ac. Sci. Paris*, t. 134, 1561–1564 (1902).

[36] Fredholm, I., Sur une classe de transformations rationnelles, *C. R. Ac. Sci. Paris*, t. 134, 219–222 (1902).

[37] Fredholm, I., Sur une classe d'équations fonctionnelles, *Acta Math.* 27, 365–390 (1903).

[38] Freudenthal, H., Operatorenrechnung–von Heaviside bis Mikusinski, *Überblicke Mathymatik* 2, 1969, Bibliographisches Institut Mannheim/Wien/Zürich. See also: *Simon Stevin*, 33 (1959).

[39] Gauss, C. F., *Disquisitiones Arithmeticae*, Leipzig 1801 (= *Werke*, Erster Band, Göttingen 1863 = *Recherches arithmétiques*, Paris 1807.

[40] Gelfand, I. M., Normierte Ringe, *Mat. Sbornik* N.S. 9, 3–24 (1941).

[41] Godement, R., *Introduction aux travaux de A. Selberg*, Sém. Bourbaki no. 144, 1–16 (1957).

[42] Gram, J. P., Ueber die Entwicklung reeller Funktionen in Reihen mittelst der Methode der kleinsten Quadrate, *J. de Crelle*, t. xciv, 41–73 (1883).

[43] Grassmann, H., *Die lineale Ausdehnungslehre, ein neuer Zweig der Mathematik, dargestellt und durch Anwendungen auf die übrigen Zweige der Mathematik, wie auch auf die Statik, Mechanik, die Lehre vom Magnetismus und die Krystallonomie erläutert*, Leipzig 1844.

[44] Grassmann, H., *Die Ausdehnungslehre, vollständig und in strenger Form bearbeitet*, Berlin 1862.

[45] Grassmann, H., *Hermann Grassmanns gesammelte mathematische und physikalische Werke.*

   a) *Ersten Bandes erster Theil: die Ausdehnungslehre von 1844 und die Geometrische Analyse*, Leipzig 1894.

   b) *Ersten Bandes zweiter Theil: die Ausdehnungslehre von 1862*, Leipzig 1896.

[46] Grassmann, H., Die Mechanik nach den Principien der Ausdehnungslehre, *Math. Ann.* 12, 222–240 (1877).

[47] Grassmann, H., Die neuere Algebra und die Ausdehnungslehre, *Math. Ann.* 7, 538–548 (1874).

[48] N.N., Hermann Grassmann, sein Leben und seine mathematisch-

physikalischen Arbeiten, *Math. Ann.* 14, 1–45 (1879).

[49] Grassmann, H., *Geometrische Analyse, geknüpft an die von Leibniz erfundene geometrische Charakteristik*, Leipziger Preisschrift1847

[50] Gunning, R. C. and H. Rossi, *Analytic functions of several complex variables*, London etc. 1965.

[51] Hadamard, J., Sur les opérations fontionnelles, *C. R. Ac. Sci. Paris*, 1903 (= *Oeuvres* t. 1, p. 405).

[52] Hadamard, J., *Leçons sur le calcul des variations*, Paris 1910.

[53] Hahn, H., Über Folgen linearer Operationen, *Monatshefte für Math. und Physik*, 32, 3–88 (1922).

[54] Hahn, H., Über lineare Gleichungssysteme in linearen Räumen, *J. für reine und angew. Math.* 157, 214–229 (1927).

[55] Hamilton, W. R., *Lectures on quaternions*, Dublin 1843.

[56] Hankel, H., *Theorie der complexen Zahlensysteme*, Leipzig 1867.

[57] Hankel, H., Untersuchungen über die unendlich oft oszillierenden und unstetigen Funktionen, *Math. Ann.* 20, 63–112 (1882); in B. Bolzano, *Reelle Wurzeln*, Ostwalds Klassiker no. 153, Leipzig 1905.

[58] Hausdorff, F., *Grundzüge der Mengenlehre*. Leipzig 1914.

[59] Hausdorff, F., *Mengenlehre*, Leipzig 1944.

[60] Hawkins, Th., Hypercomplex Numbers, Lie groups, and the Creation of Group Representation Theory, *Archive for History of Exact Sciences* 8, 243–287 (1971).

[61] Hellinger, E., Hilberts Arbeiten über Integralgleichungen und unendliche Gleichungssystemen, in *David Hilbert Gesammelte Abhandlungen* III, 94–154 (1935).

[62] Hellinger, E. and O. Toeplitz, Grundlagen für eine Theorie der unendlichen Matrizen, *Nachr. Ges. Wiss.*, Göttingen 1906.

[63] Hellinger, E. and O. Toeplitz, Integralgleichungen und Gleichungen mit unendlich vielen Unbekannten, *Enzyklopädie der Math. Wissenschaften* II C 13, 1135–1601 (1923–1927).

[64] Helly, E., Über lineare Funktionaloperationen, *Sitzungsberichte der Wiener Akademie der Wissenschaften Math. Nat. Klasse*, Bd. 121 II A1, 265–297 (1912).

[65] Helly, E., Über Systeme linearer Gleichungen mit unendlich vielen Unbekannten, *Monatshefte für Math. und Physik*, 31, 60–91 (1921).

[66] Hilbert, D., *Grundzüge einer allgemeinen Theorie der linearen*

Integralgleichungen, Leipzig 1912.

[67] Hill, G. W., On the part of the motion of the lunar perigee which is a function of the mean motions of the sun and moon, *Acta Math.* 8, 1–36 (1886).

[68] Hölder, O., Über einen Mittelwertsatz, *Göttinger Nachrichten* 1889, 38–47.

[69] Hyers, D. H., Linear topological spaces, *Bull. Amer. Math. Soc.* 51, 1–21 (1945).

[70] Julia, G., *Introduction mathématique aux théories quantiques* I, II, Paris 1938.

[71] Kellogg, O. D., *Foundations of Potential Theory*, Berlin 1929[1], 1967[2].

[72] Klein, F., *Vorlesungen über die Entwicklung der Mathematik im 19. Jahrhundert* I, Berlin, 1926; II, Berlin, (1927).

[73] Kneebone, G. T., *Mathematical logic and the foundations of mathematics*, London 1963.

[74] Koch, H. von, Sur une application des déterminants infinis à la théorie des équations différentielles, *Acta Math.*, t. 15, 53–63 (1891).

[75] Koch, H. von, Sur les déterminants infinis et les équations différentielles linéaires, *Acta Math.*, t. 16, 217–295 (1892–1893).

[76] Kolmogoroff, A., Zur Normierbarkeit eines allgemeinen topologischen linearen Raumes, *Studia Math.* 5, 29–33 (1934).

[77] Kuratowski, C., *Topologie* II, Warszawa 1952.

[78] Lagrange, J. L., *Théorie des fonctions analytiques*, Paris 1797.

[79] Laguerre, E., Sur le calcul des systèmes linéaires, Extrait d'une lettre adressée à M. Hermite. *Oeuvres de Laguerre*, t. 1, 221–267, Paris 1898.

[80] Lalesco, T., *Introduction à la théorie des équations intégrales*, Paris 1912.

[81] Lie, S., *Geometrie der Berührungstransformationen dargestellt von Sophus Lie und Georg Scheffers*, Band I, Leipzig 1896.

[82] Lie, S., Zur analytischen Theorie der Berührungstransformationen, *Gesammelte Abhandlungen* III, *Abhandlungen zur Theorie der Differentialgleichungen*, *Erste Abteilung* 96, Kristiania, Leipzig 1922.

[83] Lie, S., *Theorie der Transformationsgruppen* III, Leipzig 1930.

[84] Liouville, J., Sur le développement des fonctions ou parties de fonctions en séries dont les divers termes sont assujettis à satisfaire

à une même équation différentielle du second ordre contenant un paramètre variable, II, *J. de Math.* (1) 2, 16–35 (1837).

[85] Loomis, L. H., *An Introduction to Abstract Harmonic Analysis*, New York etc. 1953.

[86] Mikusinski, J., *Operatorenrechnung*, Berlin 1957.

[87] Möbius, A. F., *Werke* I, Der baryzentrische Calcul, 1–388, Leipzig 1885.

[88] Monna, A. F., *Analyse non-archimédienne*, Berlin 1970.

[89] Monna, A. F., The Concept of Function in the 19th and 20th centuries, in Particular with Regard to the Discussions between Baire, Borel and Lebesgue, *Archive for History of Exact Sciences* 9, 57–84 (1972).

[90] Moore, E. H., *Introduction to a form of General Analysis, Lectures delivered september 1906*, New Haven 1910.

[91] Moore, E. H., *General Analysis*, Part I, Philadelphia 1935.

[92] Naimark, M. A., *Normed rings*, Groningen 1959.

[93] Neumann, C., *Untersuchungen über das logarithmische und newtonsche Potential*, Leipzig 1877.

[94] Neumann, J. von, Eigenwerttheorie hermitescher Funktionaloperatoren, *Math. Ann.* 102, 49–131 (1930).

[95] Neumann, J. von, On complete topological spaces, *Trans. Amer. Math. Soc.* 37, 1–20 (1935).

[96] Peano, G., *Calcolo geometrico secondo l'Ausdehnungslehre di H. Grassmann preceduto dalle operazioni della logica deduttiva*, Torino 1888.

[97] Peano, G., Sur les systèmes linéaires, *Monatsh. für Math. und Physik* V, 136 (1894).

[98] Peano, G., *Opere scelte*, Vol. II, Roma 1958, Vol. III, Roma 1959.

[99] Pincherle, S., *Le operazioni distributive e le lore applicazioni all'analisi*, Bologna 1901.

[100] Pincherle, S., Funktionaloperationen und Gleichungen, *Encyklopädie der Math. Wissenschaften* II A 11, 761–817 (abgeschlossen 1905).

[101] Pincherle, S., *Opere scelte* I, II, Roma 1954.

[102] Poincaré, H., Sur les déterminants d'ordre infini, *Bull. Soc. Math. France* 14, 77–90 (1886).

[103] Poincaré, H., La méthode de Neumann et le problème de Dirich-

let, *Acta Math.* 20, 59–142 (1897).

[104] Reid, C., *Hilbert*, Berlin etc. 1970.

[105] Riesz, F., Sur une espèce de Géométrie analytique des systèmes de fonctions sommables, *C. R. Ac. Sci. Paris* 1907.

[106] Riesz, F., Sur les opérations fonctionnelles linéaires, *C. R. Ac. Sci. Paris*, t. 149, 974–977 (1909).

[107] Riesz, F., Untersuchungen über Systeme integrierbarer Funktionen, *Math. Ann.* 69, 449–497 (1910).

[108] Riesz, F., Sur certains systèmes singuliers d'équations intégrales, *Ann. Ec. Norm. Sup.* (3), t. 28, 33–62 (1911).

[109] Riesz, F., *Les systèmes d'équations à une infinité d'inconnus*, Paris 1913.

[110] Riesz, F., Über lineare Funktionalgleichungen, *Acta Math.* 41, 71–98 (1918).

[111] Saks, S., Sur les fonctionnelles de M. Banach et leur application aux développements des fonctions, *Fund. Math.* x, 186–196 (1927).

[112] Schmidt, E., Zur Theorie der linearen und nichtlinearen Integralgleichungen I, *Math. Ann.* 63, 433–476 (1907); II, *Ibid.* 64, 161–174 (1907).

[113] Schmidt, E., Über die Auflösung linearer Gleichungen mit unendlich vielen Unbekannten, *Rend. Circ. Mat. Palermo*, 25 53–77 (1908).

[114] Shohat, J. A. and J. D. Tamarkin, The problem of moments, *Am. Math. Soc.* 1963.

[115] Sikorski, R., On the Carleman determinants, *Studia Math.* 20, 327–346 (1961).

[116] Sonine, N., Sur la généralisation d'une formule d'Abel, *Acta Math.* 4, 171–176 (1884).

[117] Steinhaus, H., Anwendungen der Funktionalanalysis auf einige Fragen der reellen Funktionentheorie, *Studia Math.* 1, 51–81 (1929).

[118] Steinitz, E., Algebraische Theorie der Körper, *Journal f. reine und angew. Math.* 137, 167–309 (1910).

[119] Stieltjes, Th. J., Recherches sur les fractions continues, *Annales de la Faculté des Sciences de Toulouse*, (1) 8 (1894), T1–122, (1) 9 (1895) A5–47.

[120] Stone, M. H., *Linear transformations in Hilbert space and their applications to analysis*, New York 1932.

[121] Volterra, V., Sopra le funzioni che dipendono da altre funzioni, I, II, III, *Rend. Lincei*, ser. IV, Vol. III, 97–105, 141–146, 153–158 (1887) (= *Opere Mat.*, t. 1, 294–314).

[122] Volterra, V., Sopra le funzioni dipendenti da linee I, II, *Rend. Lincei*, ser. IV, Vol. III, 225–230, 274–281 (1887) (= *Opere Mat.*, t. 1, 315–378).

[123] Volterra, V., Sulla inversione degli integrali definiti, *Atti Acc. Sc. di Torino*, Vol. XXXI, Nota I–IV (1896) (= *Opere Mat.*, t. 2, 216–254). Rend. Acc. Lincei, ser. 5a, Vol. V, 177–185 (1896) (= *Opere Mat.*, t. 2, 255–262).

[124] Volterra, V., Sopra alcune questioni di inversione di integrali definiti, *Annali di matematica*, t. 25 (1897).

[125] Volterra, V., Betti, Brioschi, Casorati, trois analystes italiens et trois manières d'envisager les questions d'analyse, *Comptes Rendus du deuxième Congrès International des Mathématiciens tenu à Paris du 6 Août 1900*, Paris 1902.

[126] Volterra, V., *Leçons sur les fonctions de lignes*, Paris 1913.

[127] Volterra, V., *Leçons sur les équations intégrales et les équations intégro-differentielles*, Paris 1913.

[128] Volterra, V., La teoria dei funzionali applicata ai fenomeni ereditari, *Atti Congr. intern. dei Mat. a Bologna 1928*, Vol. I, 215–232. See: *Opere Matematiche* 5, 170–186.

[129] Volterra, V., *Théorie générale des fonctionnelles*, Paris 1936.

[130] Waerden, B. L. van der, Moderne Algebra, erste Auflage, Berlin 1930.

[131] Weber, H., Lehrbuch der Algebra I, II, Braunschweig 1896[1], 1899[2].

[132] Wedderburn, J. H. M., Lectures on matrices, *Am. Math. Soc.*, New York 1934.

[133] Weyl, H., *Raum, Zeit und Materie*, Berlin 1923[5].

[134] Wiener, N., Limit in terms of continuous transformation, *Bull. Soc. Math. France* 50, 119–134 (1922).

[135] Wiener, N., Note on a paper of M. Banach, *Fund. Mat.*, t. 4, 136–143.

[136] Wiener, N., *I am a mathematician*, New York 1956.

[137] Wintner, A., *Spektraltheorie der unendlichen Matrizen*, Leipzig 1929.

[138] Yosida, K., *Functional analysis*, Berlin etc. 1968.

[139] Zaanen, A. C., *Linear analysis*, Amsterdam/Groningen 1956.

Addendum

[140] Banach, S., *Oeuvres* Vol. I, Warszawa 1967.

[141] Bishop, E., *Foundations of constructive analysis*, New York etc. 1967.

[142] Bollinger, M., Geschichtliche Entwicklung des Homologie-begriffs, *Archive for History of Exact Sciences* 9, 94–170 (1972).

[143] Cayley, A., Chapters in the analytical geometry of (n) dimensions, *Cambridge Math. Journal*, Vol. IV (1843), 119–127 (= *Mathematical papers*, Vol. 1, 55–62).

[144] Coolidge, J. L., *A History of Geometrical Methods*, Oxford 1940.

[145] Monna, A. F., Hermann Hankel, *Nieuw Archief voor Wisk.* (3), 21, 64–87 (1973).

[146] Pincherle, S., Mémoire sur le calcul fonctionnel distributif, *Math. Ann.* 49, 325–382 (1897).

[147] Stolzenberg, G., *A Critical Analysis of Banach's Open Mapping Theorem*, Northeastern University, USA

Index